Inquiring into Being

SUNY series in Ancient Greek Philosophy

Anthony Preus, editor

Inquiring into Being

Essays on Parmenides

Edited by
COLIN C. SMITH

Published by State University of New York Press, Albany

© 2025 State University of New York

All rights reserved

Printed in the United States of America

No part of this book may be used or reproduced in any manner whatsoever without written permission. No part of this book may be stored in a retrieval system or transmitted in any form or by any means including electronic, electrostatic, magnetic tape, mechanical, photocopying, recording, or otherwise without the prior permission in writing of the publisher.

Links to third-party websites are provided as a convenience and for informational purposes only. They do not constitute an endorsement or an approval of any of the products, services, or opinions of the organization, companies, or individuals. SUNY Press bears no responsibility for the accuracy, legality, or content of a URL, the external website, or for that of subsequent websites.

EU GPSR Authorised Representative:
Logos Europe, 9 rue Nicolas Poussin, 17000, La Rochelle, France
contact@logoseurope.eu

For information, contact State University of New York Press, Albany, NY
www.sunypress.edu

Library of Congress Cataloging-in-Publication Data

Name: Smith, Colin C., 1981– editor.
Title: Inquiring into being : essays on Parmenides / edited by Colin C. Smith.
Description: Albany : State University of New York Press, [2025] | Series: SUNY series in ancient Greek philosophy | Includes bibliographical references and index.
Identifiers: LCCN 2024029400 | ISBN 9798855801347 (hardcover : alk. paper) | ISBN 9798855801354 (ebook) | ISBN 9798855801330 (pbk. : alk. paper)
Subjects: LCSH: Parmenides.
Classification: LCC B235.P24 I57 2025 | DDC 182/.3—dc23/eng/20241105
LC record available at https://lccn.loc.gov/2024029400

Contents

Acknowledgments   vii

Introduction   1

Section I: Proems

1. Unifying the Poem: A Divine-Modal Reading   13
   Jeremy C. DeLong

2. Parmenides's Poem as Initiation   43
   Mary Cunningham

3. Olympus as Hades: Plato and the Homeric Parmenides   59
   Alex Priou

Section II: Truth

4. Parmenides's Fragment 2 and the Meaning of Einai   81
   Colin C. Smith

5. The Veridicality of Noein and the Particularity of Noos in Parmenides's Poem and the Continuity Between Parmenides's, Homer's, and Hesiod's Usages   103
   Paul DiRado

6   Noein and Einai in the Poem of Parmenides   125
    Michael Wiitala

7   How Many Roads?   141
    Matthew Evans

8   Revelation and Rationality in Parmenides's Fragment 7   163
    Jenny Bryan

9   Signposts for the Study of Nature: Parmenides's Fragment 8   179
    Eric Sanday

Section III: Doxa

10 Parmenides's Doxa and the Norms of Inquiry: A Case Study
    of the Fragments on Astronomy   209
    Sosseh Assaturian

11 The Essential Role of the Doxa in Parmenides's Teaching   231
    Jessica Elbert Decker

12 Fragment 18 Revisited   251
    Joseph B. Zehner Jr.

Works Cited   277

List of Contributors   289

Index   291

Index Locorum   299

Acknowledgments

Many people and entities helped shape this book in various ways throughout its four-year development. These include supportive scholars like Edward Avery-Natale, Mitchell Miller, Mitzi Lee, Jackie Murray, David Bradshaw, Thomas Tuozzo, Christopher Paone, Rose Cherubin, Barbara Sattler, and Christopher Kurfess. In April 2021, the Classics track of the Kentucky Foreign Language Conference provided virtual space in which we developed several chapters. The SUNY Press editorial staff did much to help this volume actualize its potential, and the Press's anonymous readers offered many valuable insights and much welcomed encouragement. We are very thankful for these contributions, along with those of many others whom it simply is not possible to mention by name.

The editor gives sincere thanks to all eleven chapter contributors for taking part in this collective journey toward the limits of what our spirits could bear. He is deeply grateful to Brittany Frodge for wide-ranging support—and, most of all, philosophical support—throughout all stages of the book's development. Any remaining errors should be attributed to the editor.

Introduction

Colin C. Smith

The ancient philosopher Parmenides flourished sometime in the late sixth or fifth century BCE.[1] He lived in the culturally Greek settlement Elea in what is now Italy, and although later rumors and folktales have been passed down to us, no biographical details of his life are established with much certainty.[2] His sole work is a philosophical poem that was known in later antiquity as "On Nature," written in the hexameter of epic poetry like that of Homer and Hesiod. Although we are told that Parmenides published nothing beyond his poem, roughly contemporaneous evidence suggests that the contents of his text represent a larger philosophical system about which he lectured and conversed elsewhere.[3]

The poem survives in about 150 lines, likely comprising anywhere from half to a quarter of the original,[4] preserved through nineteen quotations by later authors commonly called "fragments."[5] From what we have, we can glean that Parmenides wrote in his poem a mythic tale of a young man's tumultuous journey via chariot to a limit of thinking beyond which mortals have not dared to tread. Here the traveler meets a goddess, whose address to the youth beginning at line 1.22 makes up the bulk of the poem. The goddess promises insight into both the ultimate nature of true reality and the deceptive, errant mortal discourse about the cosmos (1.28–30), uncovering the ways in which mortals have gone astray in their opinion. The truth (ἀλήθεια) that she offers seems to concern *being itself*, indicated through her use of the third-person singular "is" (ἔστι or ἔστιν) from the verb "to be" (εἶναι) without complementing subjects or predicates. In speaking the bald "is" alone, the goddess lays bare what she calls the "is" that "cannot not be" (2.3), while furthermore asserting that its opposite, "is not," "necessarily must not be" (2.5). The goddess evidently holds that this insight is deductively available to thought, although mortals have missed it, and that it in fact bears some close relationship to thinking itself (3.1). After further unraveling these philosophical realizations through a series of abstract "signs" (σήματα, 8.2) indicating the unity and omnitemporality of that sense of being that constitutes truth, the goddess embarks upon an extended "doxastic" (i.e., of δόξα, or "opinion-based") cosmogony that she marks off as a "deceptive" (8.52) example of the kind of flawed mortal account that she calls upon the youth to reject. Although she disparages it, her cosmogony nevertheless includes numerous insights into astronomy and physics: for instance, she advocates for heliophotism, or the view that the moon is illuminated by the light of the sun, and the identity of the morning star and evening star. These insights could very well mark breakthroughs in the history of science.[6]

While there is much that is astounding about Parmenides's poem, we should especially note the profound dissemination of its influence. Arguably, every historical era of Western philosophy has included major thinkers indebted to the Parmenidean insights. In antiquity and the Middle Ages, these included the Eleatic school of Zeno and Melissus that counted Parmenides as its founder, through Plato's and Aristotle's interests in a stable reality underlying perception, and the later Neoplatonists seeking insight into The One. Parmenidean thought variously reemerged in modern philosophy, including via the Cartesian cogito as deductive insight into the necessity of individuated instances of being entailed by thought, Spinozistic monism, the Kantian distinction between noumenon and phenomenon, and the Hegelian quest for absolute knowing and the absolute Geist that is its target.

Parmenides's influence has continued into twentieth- and twenty-first-century philosophy, divided as it has been into allegedly separate continental and analytic streams. We see this, for example, through Parmenides's dual influences upon European thinkers like Martin Heidegger and Jacques Derrida, and Anglo-American metaphysicians like W. V. O. Quine and Bertrand Russell, all of whom took interest in the Parmenidean notion of being.[7] Parmenides furthermore has been counted both among the earliest formative systematic logicians and the most influential mystics inquiring prophetically into the nature of transcendence.[8] He is a key figure in the formation of the Western philosophical tradition, and also one main bridge between Western and Eastern thinking roughly at the time of their historical formations.[9] Parmenides's work furthermore is among the very first concerning theoretical metaphysics via the apparent indication of the "being" that is both everywhere and nowhere, and yet his status as an applied thinker practicing astronomy and physics who was responsible for numerous scientific breakthroughs in our understanding of natural phenomena is well documented. (We could also add that speculation abounds concerning his active practice as a physician.)[10] This dissemination of influence positions Parmenides as above, or even somehow structurally constitutive of, the many oppositional dyads through which the history of philosophy is typically understood: these include the analytic-continental, logical-mystical, Western-Eastern, and theoretical-practical. Thus, Parmenides's thinking is, in a sense, superordinate to the differences and dualities through which we tell the story of our intellectual lineage.

Nevertheless, this unifying set of philosophical insights is positioned within a series of tensions and oppositions. Perhaps most immediately, this includes Parmenides's aim to have the narrating goddess uncover what no other mortal has uncovered and argue singularly against all others, apparently including all those renowned for wisdom that had come before him.[11] We must also note the historically recognized dyad, posited since antiquity, to which Parmenides is *not* superordinate: that is, the much-discussed putative opposition between Heraclitus as an alleged philosopher of endless change and Parmenides as an alleged philosopher of changeless stability.[12] These tensions expand through the various "parricides" against "Father" Parmenides (Plato, *Sophist* 241d) apparently taken up at least from Empedocles's critical engagement, onward into Plato, and continuing into the modern era in which Parmenides has been criticized for his allegedly fallacious thinking and commitment to naive metaphysical views.[13] Thus, the history of Parmenidean philosophy and its reception is one of unity, opposition, and the resultant need to resolve such an apparent tension.

Furthermore, it is by no means clear that we have understood the meaning of Parmenides's poem. Certainly the interpretations to which it has given rise have been hotly contested; Parmenides has been taken to be a radical monist who believes in the "existence" of only one thing,[14] and one who denounces the reality and value of the world that we perceive.[15] Conversely, many have defended the opposite claims, that is, that he is not especially interested in "existence"[16] and that his account entails no such denunciation.[17] The history of Parmenidean scholarship is thus riddled with disunifying debates about what aims to be an account of unity.

Far from indicating a thoroughgoing inscrutability, I suggest that these debates illustrate the perennial value of the Parmenidean insights and our engagement with them. Heidegger, himself a divisive figure in the history of philosophy, speaks well to this value, writing: "The dialogue with Parmenides never comes to an end, not only because so much in the preserved fragments of his 'Didactic Poem' still remains obscure, but also what is said there continually deserves more thought. It is a sign of a boundlessness which, in and for remembrance, nourishes the possibility of a transformation of destiny" (Heidegger 1975, 100–101). I cannot be certain that my own understanding of this passage coheres with Heidegger's meaning, but in any case, perhaps the "transformation of destiny" to which Heidegger refers entails a deeper understanding of our collectively shared nature attained as we think through the inquiries into being that Parmenides offers via the narrating goddess. Such thinking might help us to get beyond the merely apparent differences in thinking and being to grasp the unity that has underlay apparent difference all along, thereby clarifying the constitutive function of difference.

Recovering the Poem

The nineteen surviving fragments come from later authors writing between the fourth century BCE and the sixth century CE who likely had access to original copies of the poem. There is little hard evidence concerning the fit and order of the poem fragments that we have, though we can be reasonably sure of a few things at least. Sextus Empiricus tells us that the thirty-two-line quote he gives at *Against the Mathematicians* 7.111–114 (213, 4–217) is the opening of the poem, and his claim is entirely plausible. In this quotation now known as Fragment 1, we find the story of the young traveler's journey to the limit and encounter with the goddess who (apparently) gives voice to all that follows in the poem, promising the youth separate accounts of truth and opinion. Fragment 1 thus acts as a setup for what follows in the other fragments.

Fragment 8, which is the longest single excerpt at sixty-one lines, is preserved for us in full in Simplicius's sixth century CE *Commentary on the Books of Aristotle's Physics* (145.1–146.25). Writing nearly a thousand years after the poem's composition, Simplicius also bemoans the increasing "rarity" (σπάνιν, 144.29) of the copies of Parmenides's text then in circulation, and we indeed have no evidence that anyone after Simplicius had access to an original copy.[18] In any case, lines 50 through 52 of Fragment 8 seem to indicate a shift in the narrative: here the goddess states that she has concluded the "reliable account and thought" (πιστὸν λόγον ἠδὲ νόημα, 8.50) about "truth" (ἀλήθεια, 8.51) and turns now toward "opinion" (δόξα, 8.51) and its deceptive arrangements (8.51–2), the account of which follows. Because this dichotomy seems to point back to the duality between truth and opinion at issue from Fragment 1 (particularly evident in the contrast between ἀληθείης at 1.29 and δόξας at 1.30) and about which the goddess promises to give separate accounts, we likely should take lines 8.50–52 to be a structural hinge in the poem that concludes the first account, on truth, and begins the second, on opinion.

This leaves us with seventeen further fragments, none longer than nine lines, that must be speculatively ordered according to a basic structure that begins with Fragment 1 and features Fragment 8 as a central dividing point. In his *Fragmente des Parmenides* (1795), Georg Gustav Fülleborn offered an influential ordering of these into sections on truth and opinion, respectively, on the hypothesis that these fragments concern one or the other and that the poem divides into these two sections. Since then, these two sections have gone by various names, including the common "Way of Truth" and "Way of Opinion," as well as "Truth" and "Opinion," "Alētheia" and "Doxa," and so forth. A version of such a structure became orthodox through its general adoption, with emendations, in the influential Hermann Diels volume *Die Fragmente der Vorsokratiker*. This volume is best known in its later revisions by Walther Kranz, called simply "Diels-Kranz" (or DK), which constitutes what we now call the "traditional" ordering of the fragments into three sections: the Proem (Fragment 1), Fragments 2 through 8.51 on truth, and Fragments 8.52 through 19 on opinion.

There is much that is compelling about such an approach to ordering the fragments, as it coheres plausibly with the structure indicated in Fragments 1 and 8 while offering context for those remaining. But we should also note that it faces many potential objections. For starters, dividing the poem into two sections on truth and opinion nevertheless yields *three* sections since their setup in Fragment 1 cannot fit into either section! (Indeed, this will be the first of many challenges to posited binaries that we will encounter in this volume.) In any case, such a fragment ordering must answer to further potential objections, including many concerning the ambiguities—of which Parmenides himself was surely aware—of the texts' contents and its numerous opaque assertions.

While the ordering of the fragments is thus an open question, the authors in this volume do little here to challenge the traditional account. Because our aims are largely philosophical, we have generally sought here to consider the insights themselves and not the history of their textual ordering. Nevertheless, worth mentioning at the outset is that there has been much valuable scholarship in the twenty-first century entailing calls to reassess the traditional ordering; interested readers are directed in particular toward recent work by Néstor-Luis Cordero, Christopher Kurfess, Jeremy C. DeLong, and William Altman.[19] In any case and for the sake of standardization, we have generally assumed the traditional ordering and done our best to refer throughout to the three recognized sections as the Proem, Truth, and Doxa.

About This Volume

This book volume is intended to review, complement, and contribute to recent work on Parmenides. It includes new research from scholars working within and among various interpretive traditions, including analytic and continental philosophy, classics, political theory, and literary theory.

The volume is divided into three sections corresponding to the traditionally recognized sections of the poem. Section I includes three table-setting proems to the study of Parmenides. In the first chapter, Jeremy C. DeLong offers the material for a new, unified understanding of the poem's three parts as together entailing a critique of traditional Greek conceptions of divinity. DeLong critically synthesizes John Palmer's recent work on the modality of being described in Truth with a reassessment of Parmenides's role in the history of theology and Xenophanes's influential critique of divinity conceived anthromorphically. DeLong argues from this synthesis that Parmenides offers a purely deductive view of necessary being as a means of rejecting traditional divinity as embodied in anthropomorphic gods. In chapter 2, Mary Cunningham considers the initiatory purification the poem enacts with respect to both ancient Greek cultural practices and early natural philosophy. Cunningham explores the poem's aspects that lie between traditional ritual and what we would now recognize as philosophy and the Parmenidean aporiai that they yield. Alex Priou speaks in chapter 3 to some aspects of Parmenides's poetic-philosophical situatedness between Homer and Plato, and the conceptual battle over the nature of nonbeing amid the three. Priou draws parallels among characters from Homer's *Iliad* and *Odyssey*, Parmenides's poem, and Plato's *Sophist* and *Statesman* to illustrate some tensions—including the savageries of parricide and cannibalism, as well as erotic longing—in which the Parmenidean insights into nonbeing are framed.

Section II includes six chapters on the fragments traditionally assimilated into Truth. The authors of its first three chapters collectively explore the notions of εἶναι ("to be"), νοεῖν (roughly "to know"), and their relation in Parmenides's poem. In chapter 4, I return to the much-discussed question concerning Parmenides's invocation of being and its meaning throughout the poem, but especially εἶναι and its cognates in Fragment 2. I argue that the modern armaments for conceiving of being in "existential," "predicative," and "identity"-marking senses are insufficient to capture the broader and more basic notion, being as such, to which Parmenides calls our thinking. Next, Paul DiRado considers the meanings of the noun νόος (roughly "mind") and verb νοεῖν in chapter 5. DiRado argues that they diverge slightly in meaning in both the epic tradition and Parmenides's poem, and that a consideration of this divergence helps us to understand central claims in Truth. Continuing the discussion in chapter 6, Michael Wiitala considers Parmenides's provocative and vexing claim via the goddess concerning the sameness of "to know" (νοεῖν) and "to be" (εἶναι). Parmenides's phrasing is grammatically ambiguous and could mean either that the same thing is there to be and to be known, or alternatively that being and knowing are the same. While previous scholars have taken sides concerning one or the other meaning, Wiitala treats the ambiguity as a resource and argues that it is best to consider both meanings to hold simultaneously. This allows us to grasp the insight to a fuller extent.

The latter half of Section II includes three chapters on further interpretive issues in Truth. In chapter 7, Matthew Evans speaks to the controversial issue of whether Parmenides has the goddess describe two routes of inquiry, as we seem to find in Fragment 2, or instead three, including the third "mixed" route that is apparently described in Fragment 6. Evans addresses the history of the debate and argues for why he takes the latter, three-route account to be stronger. Jenny Bryan speaks critically in chapter 8 to the interpretation of Parmenides as a champion of "individuated rationality" over the "irrationality" represented, for example, by his use of a goddess as a source of philosophical revelation. Bryan focuses on technical vocabulary in Fragment 7 to argue that we must not too quickly reduce Parmenidean philosophy to a thoroughgoing endorsement of individual assessment via private cognition as the ultimate source of truth. In chapter 9, Eric Sanday considers Fragment 8 and the shift it inaugurates from Truth to Doxa by turning our thinking from being itself to that which has being (τό ἐόν, or "what-is"). Sanday argues that the goddess's "signs" concerning our mortal grasp of being itself point to continua of more and less through which spatiotemporal things derive their determinate intelligibility, and through which being itself is preserved so far as possible in mortal thinking. This prepares the turn to nature in the latter portion of Fragment 8 and the poem.

Finally, Section III of the book includes three chapters on the Doxa fragments. Many in the history of philosophical commentary have wondered why Parmenides bothered to have the goddess describe the flawed mortal accounts at such length, and the authors of these three chapters speak to Doxa's scientific and philosophical import. In chapter 10, Sosseh Assaturian draws upon recent work on Parmenides's innovations in the history of science to show the value of Doxa, the empirical knowledge it yields, and the relationship between Doxastic empirical knowledge and the philosophical knowledge at issue in Truth. Assaturian uses concepts in contemporary epistemology to distinguish zetetic norms of inquiry insofar as they are domain-specific and applies these distinctions to those Parmenides has the goddess make that led to the divisions amid the traditionally recognized sections of poem. On this reading, Parmenides's goddess criticizes mortals—and particularly earlier Ionian philosophers—for mistaking objects that come to be and perish for the abstract entities that constitute reality at the highest level, and vice versa. In chapter 11, Jessica Elbert Decker speaks to the role of the Doxa in providing Parmenides's reader (or listener) an *experience* of the kinds of insights at issue elsewhere in the poem. Elbert Decker ultimately emphasizes Parmenides's insights into the *limits* of human knowledge, and the experiential methods like spiritedness and double speak for getting beyond such limits offered by the poem. Finally, in chapter 12, Joseph B. Zehner, Jr. considers Fragment 18 and its resultant view of procreation to address other aspects of the Parmenidean cosmogony in Doxa. Zehner is interested in establishing methodological parallels between Parmenides and earlier thinkers from poetic, philosophical, and medical traditions. Zehner's argument further supports recent research indicating that Parmenides challenged patriarchal and homophobic norms, rather than contributing to them, thereby saving him from charges faced since later antiquity.

Zehner's chapter concludes our contribution to the never-ending dialogue with Parmenides and our effort to inquire further into being in the Parmenidean spirit. While the contributing authors do not agree on all interpretive details, we nevertheless share the conviction that Parmenides is worthy of our perennial reflection, and that this will bring us ever closer to the kind of philosophical truth disclosed by Parmenides's goddess.

Notes

1. Writing centuries later, doxographer Diogenes Laertius (9.21–23) claims that Parmenides flourished in the 69th Olympiad (504–501 BCE), suggesting a birth year circa 540 BCE. He is depicted in Plato's dialogue that bears his name in his mid-sixties while Socrates was still a young man, suggesting a birth year circa 515 BCE.

2. Diogenes Laertius calls Parmenides "son of Pyres" (9.21), which later doxographers follow. Plutarch's *Reply to Colotes* 1126a (first or second century CE) preserves the story that Parmenides was a politically involved lawgiver in Elea, while some ancient evidence suggests that he was known in Elea as a practicing doctor (see Pugliese Carratelli 1965). Numerous sources claim that Parmenides studied under, or was otherwise influenced by, the philosopher Xenophanes (which Jeremy DeLong defends in his chapter in this volume); others suggest influence from the Pythagoreans and notably Ameinias (a seminal discussion of this influence is Cornford 1939, esp. chap. 1). Plato depicts Parmenides as romantically engaged with his student Zeno, who was about twenty-five years his junior. In his eponymous Platonic dialogue, the Parmenides character has an aura that is consistent with Socrates's description of him in the *Theaetetus* as "venerable (αἰδοῖος) and awe-inspiring (δεινός)" and of "an entirely noble depth (βάθος)" (183e–184a).

3. Diogenes Laertius (1.16), e.g., writes that Parmenides's poem is his sole work. At *Sophist* 237a, Plato has his Eleatic Stranger character declare to the young mathematician Theaetetus, "My child, Parmenides the Great, beginning when we were children and to the end would testify regularly as follows, in prose as well as in meter . . ." (my trans.), before quoting from what is now known to us as Parmenides's Fragment 7. This, along with Plato's depiction of Parmenides as a conversant in his eponymous dialogue, seems to suggest that Parmenides's poem reflects broader philosophical activity.

4. Diels (1897, 25–26) estimates that nine-tenths of Truth and one-tenth of Doxa survive, yielding a poem of over five hundred lines. Conversely, West (1971, 221 n. 3) estimates the poem to have been about three hundred lines; see Gallop 1984, 29 n. 8 for further notes about the history of this speculation.

5. In fact, we might have a twentieth fragment in the one-line, so-called Cornford Fragment that is either a misquote of 8.38 or a similar but separate

point made elsewhere in the poem. See Cornford 1935, as well as Barnes 1979, 14–16 for a skeptical response.

6. On this, see Mourelatos 2012 and 2020, Graham 2013, Bryan 2018, Rossetti 2020, and chapters in this volume by Sanday and Assaturian.

7. See Austin 2007, 53–73 on the "continental Parmenides" broadly, and Seidel 1964 on Heidegger's Parmenides. On the origins of the "analytic Parmenides" via Russell (particularly the brief mention at Russell 1945, 49–50) and G. E. L. Owen's reception of this reading, see Palmer 2009, 19–25 and 76–78.

8. For a reading of Parmenides as proto-logician, see Wedin 2014. Example readings of Parmenides as mysticist include Kingsley 1999, Marciano 2008, and Adluri 2011.

9. See, e.g., Robbiano 2017.

10. See, e.g., Pugliese Carratelli 1965.

11. Parmenides takes this motif from the epic poetic tradition; cf. Miller 1979, 13–16.

12. We need not dwell here on the question of whether such an account of this supposed opposition is in fact well founded; Nehamas 2002 is a good introduction to the issue.

13. Examples, respectively, are Lewis 2009 and Guthrie 1965.

14. See e.g., Owen 1960.

15. See e.g., Tarán 1965.

16. See e.g., Mourelatos 2008.

17. DeLong's chapter in this volume is particularly illustrative.

18. Cf. Cordero 2004, 12.

19. See Cordero 2010 and 2017; Kurfess 2012 and 2016; DeLong 2015 and 2016, and Altman 2015.

Section I
Proems

Chapter 1

Unifying the Poem

A Divine-Modal Reading

Jeremy C. DeLong

The chapters in this book tend to focus on individually interpreting the meaning of the poem's distinct parts: the Proem, Truth, and Doxa. This is common in Parmenidean scholarship, with most philosophical treatments focused primarily on an analysis of Truth, and secondarily explaining Doxa under that framework. As a result, the primary interpretative challenge for the poem tends to be framed in this way: How can the poem's two primary sections (Truth and Doxa), which seem incommensurable in numerous ways, possibly be reconciled into a cohesive and coherent narrative? What is the meaning of, and the relationship between (a) the positively endorsed metaphysical arguments of Truth, which argues for some unified, unchanging, motionless, and eternal "reality" or type of "being," and (b) the negatively presented "cosmological" Doxa, which, in direct contrast with Truth, describes a world of coming-to-be, change, and motion? I refer to this textual tension as the "T-D Paradox."

However, if the entire poem should provide a unified narrative whole, then even the most compelling solution to the T-D Paradox faces an additional challenge: adequately explaining the meaning and function of the poem's opening section (the Proem) with that same interpretative framework. And this is precisely where so many proposed solutions to the T-D Paradox tend to be lacking. The Proem winds up being ignored or even dismissed as irrelevant by philosophical treatments, likely due to its mythical (and thus allegedly nonphilosophical) content.

I propose that the real key to resolving the T-D Paradox lies in taking seriously Parmenides's deliberate inclusion of mythical content—which is explicitly pervasive in both the Proem and Doxa—and considering the implications this has for understanding the meaning of Truth. By doing so, a straightforward way to understand the entire poem as a coherent narrative whole is revealed: Truth is an explication of the necessary qualities for any divine being, as a corrective contrast to the deeply mistaken views held by mortals (Doxa). The Proem serves as an introductory foil, invoking epic mythology purposefully to introduce the poem's subject matter.

This chapter first considers John Palmer's recent and novel modal account of Parmenides's poem.[1] Though initially promising, Palmer's specific account will prove to be problematic, particularly with respect to his attempted resolution of the T-D Paradox. Second, a case for attributing "divinity" to Parmenides's philosophical subject is presented and how this can provide a superior resolution to the T-D Paradox itself. Finally, it will be argued how a "divine-modal reading" readily incorporates all sections of the poem into a cohesive and coherent unified narrative. The upshot of this approach is a novel interpretation of Parmenides's philosophical project as providing the earliest-known deductive arguments concerning divine nature as *necessary being*, and rejecting the mythopoetic tradition (e.g., Homeric and Hesiodic), in a Xenophanean vein.[2]

Palmer's Modal View

Palmer's Modal View of Truth

The presence of modal language in Parmenides has long been recognized, yet has largely escaped close scholarly attention.[3] Palmer's novel contribution is to argue that from these modal descriptions, Truth should not be understood as an account of what "being" in general (whether monistic or pluralistic), or any "fundamental being," must be like. Rather, Truth explicates the essential qualities of any "necessary being": that is, what any *necessary being* must necessarily be like, qua necessary being.

Grounds for the Modal Reading: Fragment 2

ἡ μὲν ὅπως ἔστιν τε καὶ ὡς οὐκ ἔστι μὴ εἶναι . . .
ἡ δ' ὡς οὐκ ἔστιν τε καὶ ὡς χρεών ἐστι μὴ εἶναι.

[On the one hand, that which is and is not possible not to be . . .
And on the other hand, that which is not and must not be] (2.3 and 2.5).[4]

Palmer's modal reading is largely based upon these lines from Fragment 2, wherein Parmenides initially introduces two "routes of inquiry." The first route is "that it is (what it is) and cannot not be (what it is)."[5] To be both "is" and "cannot not be" is equivalent to that which "must be." Therefore, the first route of inquiry existentially concerns that which necessarily exists (i.e., necessary being), and predicatively inquires into the nature of what such a being must be like, qua necessary being. The second way of inquiry is clearly the contrary: "that it is not (what it is) and must not be (what it is)," which can be understood as inquiring into necessary nonbeing (both existentially, as well as the predicative nature of that mode of being).[6]

These full modal descriptions never reappear in the extant fragments. However, Palmer argues this is likely due to the constraints of the poetic meter, and so Parmenides chose to rely on a type of shorthand to refer back to them.[7] This is abundantly clear from Fragment 2 itself, when the second route, necessary nonbeing, is described as impossible to think upon, and referred back to as τό . . . μὴ ἐόν (2.6–8a). Similarly, Parmenides should be taken to refer to the first route (necessary being) by the shorthand τό ἐόν.[8]

On this reading, the first way of inquiry seeks to discover the nature of that which is now, always has been, and always will be, actually existent in the world; how any thing that necessarily exists must be in its nature, given the kind of thing it is, and the mode of its existence. And the second way of inquiry seeks to examine what any thing that necessarily cannot exist in actuality must be like, given its nature. Of course, examining that which necessarily is-not results in no object being discovered, for it has not ever, and never could be, actualized. This reading thus makes good sense of the repeated injunctions by the goddess to avoid this path, as it yields nothing fruitful, and so further inquiry should be immediately abandoned.[9]

Furthermore, this reading readily explains why knowledge of these initial two routes is entirely trustworthy and stable for Parmenides: the nature of such entities must have certain essential properties that can be readily deduced (e.g., they could be eternally existent or not), given the sort of things they are.[10] Knowledge obtained from these routes of inquiry alone will be "fixed," as only the natures of these subjects is entirely unchanging. This point is especially important for understanding Fragment 3: "the same thing is for understanding and for existing."[11] Without the context that Parmenides is inquiring into necessary being and nonbeing as the modes of existence in which "fixed" understanding is alone to be found, this line can be far too easily read on highly implausible and uncharitable Gorgian lines: that whatever can be (or actually is) thought of must also exist (cf. Gorgias, *On Nature*). The modal reading readily avoids this problem. Inquiry along the second route leads to a nature that has no conceivable content or actuality. That which must exist can have conceivability and be an object for which knowledge is entirely "trustworthy," as it "does not wander." Only with respect to modalities are conceivability and existence so readily and naturally linked.

The Modal Deductions: Fragments 8.1–49

Perhaps most compellingly, the modal approach far more readily justifies the whole set of ontological deductions in Truth by establishing that "what is" is truly eternal, ungenerated and imperishable (8.5–21), a continuous whole (8.21–25), unmoved and unique (8.21–33), and perfect and uniform (8.42–49). These essential qualities are far more straightforwardly implied by the nature of any necessary being qua necessary being, while it is quite difficult to see how Parmenides would be justified in attributing all these properties to a general conception of "being" in-itself, especially in terms of any nonnecessary (i.e., contingent) being. Palmer summarizes:

> It is difficult to see what more Parmenides could have inferred as to the character of what must be simply on the basis of its modality as a necessary being. In fact, the attributes of the main program have an underlying systematic character suggesting they are meant to exhaust the logical possibilities: What Is both must be (or exist), and it must be what it is, not only temporally

but also spatially. For What Is to be (or exist) across time is for it to be ungenerated and deathless; and for it to be what it is across time is for it to be "still" or unchanging. For What Is to be (or exist) everywhere is for it to be whole.[12] For it to be what it is at every place internally is for it to be uniform; and to be so everywhere at its extremity is for it to be "perfect" or "complete." Taken together, the attributes shown to belong to what must be amount to a set of perfections: everlasting existence, immutability, the internal invariances of wholeness and uniformity, and the invariance at its extremity of being optimally shaped. What Is has thus proven to be not only a necessary but, in many ways, a perfect entity.[13]

Admittedly, many of these same qualities could be deduced on other interpretations, if the goddess provides an injunction that entirely disallows thinking of "not-being" in any way.[14] However, that Parmenides's description implies "perfection" to being at lines 8.42–9 tells against such readings.

The "perfection" of the entity under discussion includes being limited in spatial extent and uniform throughout. To think of something as spatially limited seems to require thinking of itself as "not-being" beyond its own limits. To think of something as "uniform throughout itself" also seems to imply these very same properties not existing beyond itself, which also requires thinking of "not-being." On the other hand, if Parmenides's thesis is to explicate what a (spatially extended, material) necessary being must be like, on account of its modality alone, it is perfectly reasonable to think of a (spatially extended, material) necessary being as discrete entity, and one that must possess its modal nature uniformly throughout itself but not beyond itself. Thinking of a necessary being as "what cannot not be" does not prevent thinking of not-being beyond itself, but only with reference to itself, in its own (necessary) mode of existence. Given this, Palmer's modal reading not only makes better sense of the modal constructions that appear evident in Fragment 2 but also the central deductions in Fragment 8, while furthermore showing the deep flaws in the "one cannot think of non-being at all" approach.

Palmer's Positive Doxa

Unfortunately, positing a modal account of Truth itself does not immediately indicate precisely how Doxa should be understood. What is the subject of

Doxa itself? Is the account intended positively (i.e., as recommended), or negatively (i.e., to be rejected), and to what degree? What is Parmenides's purpose for including this account, which is so clearly contrastive to the account in Truth and its relation to the learning program of Truth? To explore these questions, Parmenides's "third path of inquiry" must first be examined.

The Third Route of Inquiry

On Palmer's modal view, the route of necessary nonbeing always and immediately leads to just one destination—the conclusion that there is "nothing at all"—and thus one's knowledge of it is fixed (i.e., it does not wander). The route of necessary being, on the other hand, leads to an a priori, logically certain, and exhaustive understanding, based upon understanding necessary being qua its mode of being. Such knowledge is similarly fixed (i.e., it does not wander). Though initially implied, these two routes for inquiry from Fragment 2 should not be understood as the *only* two conceivable routes.[15]

Parmenides clearly seems to introduce a third and quite distinct route for inquiry in Fragments 6 and 7.[16] The third route is a "mixed" path, one upon which mortals, knowing nothing and depending entirely upon their senses, erroneously think "that which 'is and is not' are the same and not the same" (6.4-9). Precisely what is meant by this will need to be determined below. However, it is clear enough that mortals are inclined to be deeply confused about *something*, and following this "wandering" route results in a lack of the epistemic deductive certainty (i.e., "knowing nothing") of the initial two routes.[17] The presence of a distinct third route also seems necessary as the subject matter of Doxa. Doxa is neither about "what is" or "what is not" exclusively, but rather things that can be both ("mixed"), at different times: that is, contingent beings.

The T-D Paradox: Palmer's Resolution

Based upon the contrasting epistemic certainty between the initial two routes of inquiry and the third route, Palmer takes Parmenides's Doxa to be anticipating Plato's own epistemological distinctions. There is only the possibility for fixed, certain knowledge about necessarily unchanging things (i.e., Forms or "necessary being"). Any inquiry into contingent being (i.e., objects of experience) will then be limited to "wandering" opinions, for which no genuine and unchanging knowledge is possible. From this, Palmer offers a resolution to the T-D Paradox.

Truth is an explication of the nature of necessary being qua necessary being. Knowledge of this mode of being can be deductively certain and unchanging. Doxa can then be cast as a contrasting account of an entirely different subject: contingent being. The nature of contingent beings is that they do come to be, change, move, change their aspects (e.g., brightness and darkness), and so forth, at different times and places and from different perspectives. And thus, knowledge of them is not "fixed" but "wanders." The error of mortals (i.e., the third route) on Palmer's reading is that by relying solely upon their senses instead of reason, mortals erroneously think contingent beings are *all* that exist in the world. Mortal views are not in-themselves mistaken about contingent beings; they simply have not learned how to understand necessary being properly, and confuse the qualities of necessary being with contingent qualities. Thus, they entirely lack understanding that does not "wander," which Parmenides seeks to rectify.

While Palmer accepts proto-Platonic epistemic distinctions in Parmenides, he rejects that Parmenides anticipated Plato's ontological hierarchy (i.e., as it entails "really-real" things as opposed to "imitations").[18] That is, Palmer does not read Parmenides as arguing that the impossibility for certain knowledge about a type of thing then implies such things are any "less real." The contingent world does exist (and/or "is what it is"), and there is still value in learning what one can about it, even if that knowledge is not certain. So, even despite its epistemic inferiority, its unavoidable "wandering," Palmer avers that Doxa is Parmenides's own best attempt to explain the world of contingent being and should be considered *positively* endorsed as such. Given that the goddess herself explicates Doxa as part of the learning program of the poem, Palmer reasons that "it is apparently not altogether wrong to follow it."[19]

Unlike more "monistic" readings in which Truth refers to "all of reality," Palmer's more restricted scope for Truth (i.e., as concerning necessary being qua necessary being) avoids the world-denying or "mad" entailments of many "strict monism" interpretations that plagued much of nineteenth- and twentieth-century scholarship on Parmenides.[20] Casting the crucial distinction between the "trustworthiness" of Truth and Doxa in terms of an epistemic rather than ontological distinction also helps avoid this issue. Taking Doxa to be a positively endorsed "best possible" account of contingent being, which is not wrong about its own subject matter but merely lacks the epistemic certainty of Truth, the presence of

Doxa's contrary account to Truth is more sensible. By making Truth and Doxa about two distinct modes of being (i.e., necessary and contingent), rather than correct and incorrect beliefs about the same ontological subject, it is no longer necessary to resolve mutually exclusive accounts about the same subject. Taken together, Palmer's modal reading appears to have neatly sidestepped the T-D Paradox entirely. Yet, despite this initial promise, I demonstrate below that there are significant intertextual challenges, as well as interpretative lacunae, that result in substantive strikes against Palmer's view.

Challenges for Palmer's View of Doxa

There is no doubt that the poem treats Doxa negatively in comparison to Truth. Palmer attempts to mollify this by treating Doxa as an epistemically inferior yet "best possible" account of contingent being(s). Contra Palmer, I argue the text requires a far more negative view of Doxa. Furthermore, I will argue that Palmer's particular solution to the T-D Paradox cannot possibly succeed, as it is clearly inconsistent with the Doxa passages that most explicitly address the error mortals have committed. I refer to this error below as the "Naming Error."

Why Parmenides Cannot Be Endorsing Doxa

Doxa is first described as having "no true trust" in the Proem (1.30). Palmer's account of Doxa concerning contingent (and thus changeable) things, lacking a "fixed" certainty, would seem to satisfy this. His view can also be taken to satisfy the stronger assertion that the account in Doxa is "deceptive" (8.32). Since contingent beings and their qualities are changeable, claims about them can be deceptive, in that they can mislead the viewer or knower and fail to provide complete and unchanging understanding.[21] Doxa could also be understood as deceptive on Palmer's view in the sense that by only having experience of contingent being, mortals are misled to believe this to be the only kind of being there is, thereby failing to recognize (or even denying) the modality of necessary being entirely or that anything actualizes that mode. The related claim that mortals "know nothing" (6.4) can be understood along similar lines.[22]

None of these points are in direct tension with Palmer's reading of Doxa as an account of contingent beings that Parmenides positively endorses as a "best possible" account of such. However, if the poem intended Doxa to be endorsed in any way, the goddess should not then so explicitly and universally enjoin the anonymous male youth (κοῦρος; often referred to as the "philosopher-youth") from following the "mixed path" of wandering inquiry, as she does in both Fragment 6 and 7. Here is Fragment 6:

> It is necessary to say and understand that this is (necessary)
>   being, for it is for being,
> And nothing ("necessary non-being"?) is not. These things I
>   urge you to consider.
> For first **I restrain** you from this path,[23]
> And thereafter from this other path, the one which mortals,
>   knowing nothing
> wander, two-headed. For a lacking in their breasts
> guides the wandering mind. And they are borne along
> deaf and blind alike, wondering, a tribe without judgment,
> to whom to be ("come-to-be") and not to be are thought to
>   be the same
> and not the same. And it is a backwards-turning journey for
>   all of them.[24]

The goddess thus clearly seems to "restrain" the youth from following the "mixed path of mortals" (Doxa), describing this route as an egregious error resulting from mortal ignorance. While Palmer attempts to offer a creative emendation at line 6.3, removing the goddess's "restraint" from the "mixed path" in favor of a more neutral "programmatic outline" of the goddesses upcoming discussion, this creative solution is far from convincing.[25] For even if accepted, this emendation still fails to provide the positive "best possible" endorsement of Doxa that Palmer's reading requires. And it does nothing to counter the quite negative description of the "mixed path" in the ensuing lines, which Fragment 7 then repeats and develops.[26]

Palmer also argues that Parmenides endorses the account in Doxa based on lines 8.60–61, where the goddess seems to explain her purpose for providing Doxa as an account:

> To you I relate this ordering of things, fitting and entire,
> So that no understanding of mortals may ever surpass you.
>   (8.60–61)[27]

Palmer argues that the goddess's intention here is to provide the philosopher-youth the very best account of contingent being possible—a truly superior cosmology, describing the origins, nature, and operation of the world's mutable entities—which no other mortal could best. It is not certain that, by providing this account, the goddess is "granting her authority" (and thus "reliability") to it, as Palmer would have it.[28] Furthermore, didactically explicating an erroneous account indicative of those held by ignorant mortals, one that is clearly denigrated and explicated as what *not* to follow through the poem, is also perfectly consistent with the goddess's nature, and her express purpose for offering the Doxa at Fragment 8.60.[29]

More troubling, the goddess promising the "very best account of all the contingent world" is a rather strong and strange claim for which Parmenides himself would need some justification. First, there is the poem's consistent denigration of the "mixed path" (i.e., Doxa), as noted above. Second, unlike Truth, there are no epistemic criteria or modes of argumentation offered for determining superior, as opposed to inferior, accounts of contingency. Hence, it is not clear on what grounds any account would be determined superior. Third, there is the tension between the purported lesson of Doxa on Palmer's interpretation—that contingent objects lack any "fixed understanding"—and any attempt by a mortal (e.g., Parmenides) to provide a "superior account" of such.

Palmer suggests Parmenides also had interests in empirical investigations, and quite speculatively suggests that he was likely incorporating these in the Doxa while believing them to be correct and "superior" to prior accounts.[30] There is no reliable extant evidence for this, and even if Palmer were correct in that Parmenides was also a "natural philosopher" who aimed to incorporate his own empirical observations (e.g., astronomical or biological) into his poem, this would also not be sufficient. No such empirical studies would be grounds to claim "superiority" regarding the majority of the content in Doxa for which there is extant evidence, such as the origins of the cosmos, the creative activities of divine beings, and so forth.[31] Why should Parmenides think his own theogony superior to Hesiod's, or that of any other poet?[32] Like most other commentators, Palmer generally overlooks the pervasive mythological content in Doxa (and the parallels in the Proem) in favor of casting it as naturalistic cosmology. Yet the presence of this content substantially undercuts viewing it as Parmenides's own best-possible account of contingent being.

These challenges, even if not decisively fatal to Palmer's reading, should show that this interpretative framework is quite speculative, unconvincing, and often at odds with the text. Instead, the textual evidence far better supports taking Doxa as didactically negative, functioning as a representation of the crucial errors of mortals about the subject of Truth itself for which the corrective account in Truth has been developed. The reasons for this will be clarified further in the discussion of the "Naming Error" of mortals below, which does seem to be a fatal blow to Palmer's proposed solution to the T-D Paradox.

The Naming Error

Palmer has adopted Ebert's "restoration" of lines 8.34–41, placing them instead between lines 8.52 and 8.53.[33] While contestable, this restoration can be charitably granted and Palmer's view evaluated on his own reconstruction and preferred translation. The decisive blow to Palmer's interpretative framework will persist either way.[34] Consider lines 8.38b–59:

> **To it** all things have been given as names   8.38b
> all that mortals have established in their conviction
>   that they are genuine,
> both coming to be and perishing, both being and not
> and altering place and exchanging brilliant color.   8.41
> For they fixed their minds on naming two forms,   8.53
> one of which it is not right to name, wherein they
>   have wandered astray:
> but they distinguished opposites in form, and assigned
>   them marks
> distinct from one another, for the one, the ethereal
>   flame of fire,
> being gentle, most light, every way the same as itself,
> yet not the same as the other; but that one is in itself
> the opposite, dark night, dense in form and heavy.   8.59

The interpretative crux lies in the understood subject of lines 8.38b–41, which is uncontroversially referring back to the subject explicated in Truth and thus must be necessary being for Palmer.[35] It is "to it" (τῷ), *that* very same entity (τό ἐόν), which mortals are here said to have "given as names" all the qualitative attributions listed: coming to be, perishing, altering positions, and so forth.[36] In this context, that mortals "in their conviction" believe these names to pick out something "genuine" (ἀληθῆ) with respect to the subject of Truth clearly implies that mortals are in error to do so. That is, these qualities do not correctly specify the nature of Truth's subject. This is *the* central claim about mortal errors by the goddess, and it must be convincingly met. Lines 8.53–59 then picks up on the explication of the error and how it occurred. The Naming Error of mortals—which should be understood as the reason for providing the entire Doxa, based upon Fragment 19—*requires* an account of how mortals are wrong about the subject of Truth itself.

Yet, undeniably, to specify the nature of Truth's subject incorrectly, mortals must have some conception of that relevant subject (or type of entity). Palmer himself recognizes this difficulty, asking, "How can mortals describe or misconceive What Is [necessary being] when they in fact have no grasp of it?"[37] The answer is, of course, that they cannot. The Naming Error does not just require that mortals be wrong about *something* tangentially related to Truth. Instead, the Naming Error requires that mortals actually *misunderstand* the nature of the subject of Truth itself (i.e., necessary being, on the modal reading), with Doxa as an account that explains or demonstrates this misunderstanding. Thus, on a modal reading, mortals cannot be "unaware of" necessary being(s), as Palmer would have it; rather, they must be aware of, but fundamentally *misunderstand,* necessary being(s).

Palmer must then explain how mortals could have erred by thinking only contingent being existed but also described necessary being incorrectly when mortals purportedly never conceived of that mode of being as a possibility. This is his explanation, in direct answer to his question quoted above: "Apparently because mortals are represented by the goddess as searching, along their own way of inquiry, for trustworthy thought and understanding, but they mistakenly suppose that this can have as its object something that comes to be and perishes, is and is not (what is), and so on. Again, the goddess represents mortals fixing their attention on entities that fall short of the mode of being she has indicated is required of a proper object of thought."[38] There are no grounds in the text for imputing upon mortals a search for "trustworthy thought and understanding" *about* necessary being itself. More importantly, this answer is no answer at all to the clear requirements of the Naming Error. Mortals erroneously thinking that contingent beings can provide "trustworthy thought and understanding" may indeed be *an* error of mortals; but this is certainly not the same error as mortals thinking that the subject of Truth (i.e., necessary being) can be properly described in ways contrary to its nature (i.e., coming to be, perishing, etc.), which is precisely the error the goddess insists they commit. The required reference to Truth's subject ["to it" (τῷ)] in the Naming Error passage, on Palmer's own modal interpretation, prohibits taking Doxa as a positively endorsed "cosmology" or "best account of contingent being." It makes clear that Parmenides was not at all interested in positively recommending his own "physics," but rather explaining *how* mortals get the subject of Truth wrong.

The Naming Error further makes clear that Palmer's entire account of how mortals err, and thus his proposed resolution to the T-D Paradox, cannot succeed. The structure of the poem simply does not allow for a secondary *distinct* subject (i.e., contingent being) to be the subject of Doxa, in contrast to that of Truth's subject (i.e., necessary being). The Naming Error and the framing of Doxa require Doxa itself to be an explanation of how mortals are wrong about the nature of Truth's own subject, not merely their failing to recognize its existence or futilely seeking fixed knowledge in all the wrong places (i.e., contingent objects).

If a modal interpretation of the poem cannot be made consistent with the Naming Error, then the entire interpretative framework is at risk, including all its clear benefits for understanding Truth itself. Fortunately, there is a solution to this problem. It requires adding a descriptive quality to the subject of Truth, and thus the entire poem, that Palmer himself considered carefully but ultimately chose to reject: necessary *divine* being.[39]

Divine Elements in Parmenides

Palmer rightly considered the close parallels between Parmenides's subject in Truth and Xenophanes's greatest god, yet he ultimately rejected identifying Parmenides's aim in Truth as an attempt to define divine necessary being. In this section, I build upon the correlations Palmer himself noted and argue that Palmer's modal reading would have benefitted from pressing these close parallels further.

Generally, when an entity is conceived as "perfect," "necessary," and "eternally unchanging and self-same," it is also typically described as "divine." Were this also the case for the Presocratics, and for Parmenides in particular, it could provide a straightforward solution to the problems for Palmer's interpretation raised above. For if any necessary being is also a "divine" being, since mortals certainly have conceptions of divine beings (like those in the Greek Pantheon, etc.), Truth can be taken as an account of how divine beings should be properly conceived of, while the Naming Error can straightforwardly be read as how mortals have mistakenly described divine being(s): in anthropomorphic and mythopoetic ways.

SOCIAL-HISTORICAL CONTEXT

Given Parmenides's historical context, it would be far more surprising if Parmenides did not identify "what is" as divine in its nature. The Ionian Presocratics Thales, Anaximander, and Anaximenes all associate their fundamental archē with divinity. However, a closer contemporary to and likely influence upon Parmenides is Xenophanes, who explicitly challenges traditional mythopoetic accounts of divine beings and sets forth his own "rational theology" (see Xenophanes fragments 11–12, 14–16, 23–26, 34.) While many scholars have been skeptical of Xenophanes's influence on Parmenides, there are many reasons to consider this influence more carefully.[40]

First, the question arises as to whether it was even possible for Xenophanes to have taught or influenced Parmenides.[41] Xenophanes would have lived long enough to have encountered a mature Parmenides. He left Ionia as a young man, spending most of his life in and around Parmenides's own region of Magna Graecia in what is now Italy. It is even attested that Xenophanes wrote celebratory odes on the founding of Parmenides's hometown of Elea, in addition to his native Colophon; both might be thought of as his "homes" at different stages in life (DL *Lives*, ix, 18–20). Aristotle also attests that the citizens of Elea once consulted Xenophanes regarding a ritual sacrifice (Ar.*Rhet* II.23.1400b6). So yes, it is certainly *possible*.

In fact, it is highly likely there was a direct influence. In addition to attestations that Xenophanes was Parmenides's teacher, the philosophical parallels themselves are compelling.[42] First, there are striking metaphysical similarities. Xenophanes's God is supreme (μέγιστος), transcendent (οὔτι δέμας θνητοῖσιν ὁμοίιος οὔτε νόημα),[43] unified (οὖλος) in being and senses,[44] entirely still and unmoving (αἰεὶ δ'ἐν ταὐτῷ κινούμενος οὐδὲν . . .),[45] morally perfect,[46] and eternal.[47] Similarly, the subject of Truth is concluded at 8.3–4 to be eternal (ungenerated [ἀγένητον] and imperishable [ἀνώλεθρόν]), unified (οὖλον), unique (μουνογενές), motionless (ἀτρεμές), and perfect (τέλεστον). Truth's subject might also be described as "transcendent," in that knowledge of its nature lies beyond mortal sense-perception. Both hold the most fundamental metaphysical entities in their respective systems to be motionless, changeless, unified and complete, transcendent and (in some sense) perfect.[48]

Epistemically, there is a common concern regarding the gap between mortal and divine knowledge and whether mortals could overcome this.[49] Xenophanes is more skeptical, claiming mortals not only lack knowledge of the gods but seemingly denying such knowledge is even possible (though this does not stop him from offering his own mortal account of divinity).[50] Parmenides similarly claims mortals lack all proper understanding of Truth's subject (1.1.29–30; 6–7, 8.38b–41, 8.50–54, 8.60–2), describing them as entirely lacking in understanding and judgment (6.4–9). Not only do both offer similar descriptions of their most important metaphysical subjects, but both aver that mortals face substantial difficulty (if not impossibility) in recognizing this subject's true nature.

Palmer admits the substantial analogues between the subject of Parmenides's Truth and other Presocratic views of divinity: "One cannot ignore the fact that What Is, in [Parmenides's] system, is the analogue of the greatest god [of Xenophanes], the Sphere under Love/the divine mind [of Empedocles], and Mind in the systems, respectively, of Xenophanes, Empedocles, and Anaxagoras."[51] Palmer also grants that Xenophanes most likely had some influence on Parmenides's own views.[52] Nevertheless, Palmer ultimately rejects identifying the subject of Truth as divine for two basic reasons. First, Parmenides never explicitly refers to Truth's subject as a "god," or describes it as "divine," in the extant fragments. This is true; and given the apparent completeness of Fragment 8, such reference is unlikely to have simply gone missing were it previously present. Second, the subject of Truth possesses none of the attributes normally associated with religious entities, such as mental activity, or any active or creative ordering capacities. In contrast, Xenophanes's God possesses causal control of the cosmos through thought alone, a quality notably absent from the subject of Truth.[53]

Nevertheless, numerous ancient commentators did ascribe divinity to the subject of Truth. Aetius explicitly claims that the unmoved, limited, and spherical entity described by Parmenides *is* God.[54] Philo of Alexandria wonders how Xenophanes and Parmenides—such "divine men," *theologians* engaged in theological matters—lacked sufficient divine inspiration to compose good poetry.[55] Numerous other independent sources from a variety of perspectives and times (first to sixth c. CE) make similar attributions to Parmenides.[56] Not only do many describe Parmenides as explicating the nature of divine being itself, some even note he was critical of mortal views about the divine based on anthropomorphic and moralistic grounds. Clearly many ancient readers considered the close parallels between these thinkers, especially regarding divinity, to be worth noting. And while healthy skepticism is appropriate, it is worth paying some attention to this.[57]

While Palmer is correct to note the striking dissimilarities between these thinkers, pupils often hold substantially different views from their teachers; so, this does not demonstrate a lack of influence, especially when the substantial similarities and ancient attestation are more in favor or a close intellectual relationship. Furthermore, while the subject of Truth is never explicitly referred to as "divine," it is not as if Parmenides has ignored divinity in his poem: mythopoetical and anthropomorphic conceptions of divine beings are pervasive throughout the Proem and Doxa. And these extensive references to divine beings in the poem do demand some explanation.

Truth's Divine Bookends: Proem and Doxa

The Proem clearly offers a mythopoetical framing for the poem (1.1–32) full of references to divine beings.[58] It is a first-person account from the philosopher-youth, relating his cosmic chariot journey upon a "divine path" (1.2–3) driven along by the Heliades (Maidens of the Sun, 1.9). Starting from the House of Night, the Heliades enter into the light, leaving the darkness of Tartarus behind (1.9–10). Presumably, they collect the youth in-between, and ultimately arrive at the "aetherial" Gates of Night and Day guarded by Justice (1.11–15). Persuading Justice to let them pass, the party passes into a yawning chasm beyond (1.15–17).[59] An unnamed goddess then welcomes the youth, assuring him that no evil fate, but instead Rightness (Θέμις) and Justice (Δίκη), has delivered him to her dwelling, "far from the path of men" (1.24–28a). The goddess outlines her learning program for the youth, and from then on, the entire poem (Truth and Doxa), consists of the teachings of the goddess.

It is certain that epic hexameters in this section are largely borrowed from Homer.[60] The religious imagery of Light/Night, the House of Night, and the Gates of Night and Day closely track Hesiod's *Theogony* (Hes. *The.* 713–819). Parmenides is clearly and intentionally invoking the mythopoetic tradition here, with numerous references to certain (anthropomorphic) divine beings and locations. But for what purpose? There is no immediately obvious or explicit philosophical message here, or a way to link it directly to the message in Truth; this is likely why it is so often minimized or even ignored by philosophical commentators.[61] What has gone largely unnoticed is how the themes in the Proem are echoed in Doxa.

Note the pervasive thematic focus upon contrasting imagery of Day/Light and Night/Darkness in the Proem. The Heliades set out from dark Tartarus, veiled and in darkness, only unveiling as they "enter the light." As Daughters of the Sun, the "divine path" they traverse is most likely the diurnal path, shared by both Sun/Day and Night.[62] Though the description of the Gates of Night and Day as aethereal has suggested a heavenly location to many commentators, the mythopoetic tradition always locates these gates in dark Tartarus. This is also true for the House of Night, where Night and Day alternatively dwell.[63] As the Heliades initially are driving into the light, it should be daytime, with Night in residence, suggesting an identification for the unnamed goddess. The ambiguities of location and alternating descriptions draws a strong contrast between light/heavens and darkness/Hades.[64]

While this thematic contrast is entirely absent from Truth, there are notable parallels in Doxa. The sparse fragmentary evidence from Doxa makes constructing a unified narrative account challenging, but there are some recognizable themes. It begins with the Naming Error, which explicates how mortals have erred regarding Truth's subject, particularly by "naming" (ascribing qualities) to that subject in terms of a Light/Night duality for the entire cosmos: a duality that appears to serve as the constitutive elements of "all things" via their mixture (8.51-61, 9). The explication of this error seems to be the overarching thesis of Doxa, and it centers on this Light/Night duality and the Naming Error it entails.[65]

The ensuing passages seem to promise a forthcoming detailed account of the origin of celestial bodies like the sun, moon, stars, aether, heavens, and so forth (Frs. 10-11). Yet this cosmogonical account seems in fact to begin with a *theogonical* account instead, wherein a new primordial goddess (also anonymous) rules the entirety of the dualistic Light/Night cosmos—at least in terms of creation or birth—from which she begins to create other divine beings, starting with Eros (Frs. 12-13). There are then some hints of the promised astronomical descriptions: the moon as reflecting light from the sun, and a single-word description purportedly related to the Earth being "water-rooted" (14-15a). There is also some consideration of the relationship between the mind and body (16), and discussion of animal and human procreation (Frs. 17–18). The entire Doxa is then tied together as a unified account, with a conclusion directly referencing back to the Light/Night Naming Error (19).

The pervasiveness of the Light/Night Naming Error should now stand out as the primary contextual framing for the entirety of Doxa. Lines 8.53–61 must introduce Doxa, and it does so by introducing the Naming Error. Granting that Fragment 19 is indeed properly placed as the conclusion of Doxa (as it seems to be), and given that it explicitly refers back to how mortals have "erred in naming," then the entire Doxa should be read as a development or explication of that central issue and interpreted in that contextual light. This interpretative context seems to have been missed by both ancient and modern commentators, as they tend to treat Light/Night duality introduced in lines 8.53–59 as concerning a fundamental physical dualism, that is, not divine entities. Yet the Doxa itself goes on to develop the Naming Error explicitly as a theogony or cosmogeny, belying the reductive physicalist account.

The Proem also explicates a paralleled Light/Night dualism in clear divine terms. Light is symbolized there by the Heliades themselves (i.e., the daughters of Helios/Sun), who "go into the light," steering a chariot with "flaming wheels," and arriving at "aethereal" gates. Night is symbolized by setting the narrative goddess, likely Night herself, in the House of Night, in which both Sun/Light and Night alternatively dwell, and which is beyond the Gates of Night and Day/Light within a yawning chasm in dark Tartarus. Given the paralleled Light/Night Dualism in the framing of the Proem and its clear divine associations, it would seem appropriate to infer that a similar identification is likely intended in Doxa.

Just as the Proem focuses on seeking the teaching of an unnamed spokes-goddess (the character in the poem that narrates both Truth and Doxa), the central figure of Doxa is a similarly anonymous cosmic-goddess, one who "controls all things, by mixing together male and female." This causal efficacy seems to include mixing primordial gendered opposites, like Light/Sun (male) and Night (female), as well as human and animal sexual activity. Furthermore, the first (and only extant) action explicitly attributed to this cosmic goddess is theogonic, with Eros being created as the "first of the gods." This is fitting, as Eros is a primordial force related to combination or mixture and sexual generation, as readily seen in Hesiod's *Theogony*.[66] The first and second generations of primordial divine entities (including Night and Eros) are spontaneously generated by a creative account of will, without sexual mixture, according to Hesiod. Only after

Eros has come to be does sexual union and generation occur between deities of the second generation in Hesiod's *Theogony*. More notably, Night herself is the first deity in Hesiod's *Theogony* to engage in sexual union and birth, coupling with her brother Erebus (Darkness) to generate Aether and Day (Hes. *The.* 116–25). Analogically in the Proem, the Daughters of the Sun (Heliades) first spring from the darkness of Tartarus and the House of Night, ascending to the aethereal gates prior to returning to the abode of Night. Building further on the parallels between Proem and Doxa, if the spokes-goddess herself is Night, it seems the cosmic-goddess of Fragment 12—the one who controls all sexual and generative activity—might also be so identified. The spokes-goddess might even be seen as providing an account of how mortals err in understanding her own nature.

Fragments 14–15a then do describe some aspects of the sun, moon, and earth. Focusing on this content in isolation, Doxa might seem to be a physical cosmology, yet it should not be forgotten that all the named entities normally taken to be mere physical bodies are also divine entities or locations: Aether, Ouranos, Helios (Sun), Selene (Moon), Olympus (an actual mountain, and mythical home of the gods), Astraios (stars), and Gaia (Earth). Given the preceding theogonical context of the poem, the textual evidence is against taking these to be descriptions of mere physical bodies, rather than divine beings. Further support for this reading comes from noting that the behavior of these entities is also anthropomorphized: for example, Selene always "gazes" toward the sun and "wanders" around Gaia. Carrying the theogonical context forward to Fragments 16–19 reveals that a similar divine emphasis can be consistently applied—an account of how the cosmic goddess generated and established the nature of mortal intellect and sexual reproduction—rather than any purely secular physicalist account.[67]

By recognizing how the Proem and Doxa are carefully contrived in their theistic parallels, it becomes clear that the Proem must be interpretatively significant, and directly related to the forthcoming account in Doxa. Furthermore, mapping these parallels reveals that the purportedly "Parmenidean physics" should not be considered a physics at all but rather a theogony closely related to Hesiod's, though with an entirely different purpose. This alone might help make some sense of the claims that Parmenides was engaged in theistic considerations, as any theogonical account clearly counts as engaging in divine inquiry. However, this explanation still requires a rather obvious misreading of the poem: taking Doxa to be positively endorsed.

32 | Jeremy C. DeLong

By carefully noting textual parallels between the Proem and Doxa, a striking theme is revealed: a pervasive focus on divine beings cast in anthropomorphic roles. And since Doxa is a mistaken view of mortals, this should inform us of Truth's actual aim: a correction of mortals' errant views about the nature of the divine, which in fact is always perfect and unchanging necessary being. Thus, Parmenides is best understood as providing the earliest known deductive arguments concerning divine nature, and what must necessarily follow from the essential nature of any necessary being qua necessary being, wherein any necessary being is a divine being. In the process, Parmenides ends up rejecting the mythopoetic traditional understanding of divinity in a Xenophanean vein by denying anthropomorphic characteristics to the divine. However, Parmenides was willing to go even further than Xenophanes himself and follow the reasoning to its logical conclusion. Having a perfect necessary being that is not anthropomorphic or changeable in any way requires eliminating mental activity and cosmic control. All of these require changing, or coming-to-be, and as such are inconsistent with a perfect, eternally unchanging being.

To get at Parmenides's actual theistic point, the religious content of the Proem and Doxa must now be considered together, in relation to Truth, and the poem's negative treatment of Doxa based on the Naming Error. A divine-modal reading can readily pull all these elements together.

Divine-Modal Theism and Interpretive Significance

This chapter has developed and defended the following subtheses, in an order roughly relevant to how they would be encountered and recognized when reading the poem:

1. The Proem carefully invokes the mythopoetic tradition concerning divine being.

2. A modal reading makes better sense of the deductions in Truth.

3. Truth's subject was likely strongly influenced by Xenophanes's conception of divinity.

4. Ancient Greek culture implies that we should readily identify Truth's subject as divine in nature.

5. Doxa must be negative, and a description of how mortals misunderstand the subject of Truth.

6. Doxa is likely more theogony than physical cosmology.

7. As the Proem and Doxa both contain extensive divine references and bookend Truth as introduction and counter-thesis, this extensive divine content should be directly relevant to the discussion of Truth's subject.

If these points can be granted together, they recommend ascribing divinity to Parmenides's subject in Truth and considering Parmenides's poem in a far more Xenophanean vein. With these main points, it can now be shown how a divine-modal reading can make better sense all of these subtheses, resolve the T-D Paradox, and unify the poem into a coherent narrative about one topic.

The Proem can now adequately set the stage in a more straightforward manner. While the specific story it tells is likely novel, its mimesis of and clear references to epic poetry, as well as extensive references to divine beings and locations, prepare the ancient Greek reader to hear a similar story about the gods. A goddess is then introduced as the poem's spokesperson, who promises to reveal her own divine knowledge, correcting a mortal youth's ignorance.

The spokes-goddess then proceeds to develop her "trustworthy" account in Truth of necessary being qua necessary being as a perfect and completely unchanging (form of) entity, and the qualities that any such being existing in this mode must possess. Anyone aware of earlier Presocratics's archai, especially Xenophanes's conception of divinity, could hardly fail to notice the similarities. Furthermore, the expectations for the poem's content, as established by the Proem, should still be in force. Even if a divine identification is not explicitly stated, it would be readily understood in such a way to the Greek philosophical mind.

Doxa is then presented: the wandering, untrustworthy, and ultimately erroneous view held by mortals (1.28b–30, 6, 7, 8.50–52). The Naming Error first makes clear that Doxa sets forth an example account of how mortals have erred in their understanding of Truth's own subject (and not some distinct subject, i.e., contingent being) by thinking that Truth's subject comes-to-be, perishes, both is and is not, changes and moves, and so forth (8.37b–42, 8.53–61). As established above, the theogony of the Doxa ensues, closely mimicking Hesiod's own: there are initial primordial forces (e.g., Light/Night, the cosmic-goddess) already in place. The cosmic-goddess then engages in creative activities via her mind, leading to the generation of additional divine beings (Eros first), and then other divine physical objects of the solar-system (Sun, Moon, Earth, Stars, etc.). Each generation of divine beings becomes more and more anthropomorphized in description than the previous, until the discussion moves on to more earthly (i.e., human and animal) procreation.

Doxa can finally be seen for what it truly is overall: a mythopoetic theogony and cosmogeny, describing anthropomorphized gods engaged in many activities involving creation, coming-to-be, change, motion, perishing, and so forth. These are precisely the activities and properties previously denied to necessary being in Truth. As Doxa shows how mortals have failed to understand Truth's subject properly, then on the modal reading, Doxa needs to be an account of how mortals have misunderstood the nature of necessary being. The lesson that mortals have failed to understand is that necessary being is not at all like the anthropomorphic descriptions of Doxa's theogonical cosmogeny. The implication here is that, in contrast to Doxa's theogony, Truth's subject is intended to be understood as corrective of mortal understanding of divine nature (primarily). Parmenides has simply developed and argued for this conception of divinity in terms of a necessary mode of being (secondarily).

As both the poem's bookends (i.e., Proem and Doxa) invoke the mythopoetic conception of divine nature, ascribing various anthropomorphic descriptions to divine beings, Truth can thus be understood as a corrective and contrastive account to the mythopoetic tradition in general, offering an entirely nonanthropomorphized account of divine nature instead. And based upon the Doxa's theogony, this rejection of any anthropomorphizing of divine (necessary) nature seems to extend even so far as to deny divine (necessary) nature any causally efficacious mental capacities or active role in the generation and governance of the cosmos. While the conception of the goddess as a divine spokesperson for the poem must be rejected as an inappropriate conception of divine (necessary) being, this should not be understood as an "internal inconsistency." Rather, the presence of such poetic devices should be understood as narrative vehicles, appropriate for the culture, through which to prepare the audience to receive the corrective message. Precisely such familiar mythopoetic contexts in the Proem signal to the ancient reader the poem's ultimate subject matter. Placing the corrective (though self-denying) account in the mouth of the spokes-goddess simultaneously plays upon the mythopoetic tradition's typical warrant for justification (i.e., wisdom delivered by divine beings) while also undercutting this trope by employing deductive argumentation accessible to mortals, yielding no need for divine appeals and setting forth a corrective account of divinity that leaves no room for such anthropomorphic interactions.

Such critiques aimed at Homer and Hesiod in particular, and the tendency to anthropomorphize divine nature in general, should sound very familiar. This is precisely what Xenophanes is most famous for: denouncing anthropomorphic views of the gods as "the fictions of our fathers."[68] Accepting that both were engaged in such a similar critique of mythopoetic conceptions of divinity also better explains why the divine subject in Truth so very closely mirrors Xenophanes's own conception of a supreme (only?) divine being: both are eternal, non-anthropomorphic, not engaged in immoral activities, unchanging, and motionless.[69] Together, these insights also make sense of the ancient testimonia averring a close intellectual relationship between these two thinkers.[70]

It is true that the deductions in Truth result in a complete elimination of all anthropomorphic characterizations, including mental activity, as well as causal efficacy in and/or control of the world. Parmenides's "divine" being can thus be criticized as entirely lacking in religious import.[71] Recall that these qualitative lackings are among the reasons Palmer denies attributing divinity to Truth's subject, as not adequately fitting with Greek views of divinity in general and being very different from Xenophanes's own conception. However, given their different argumentative strategies, these differences should not be so shocking. Xenophanes drew his conclusions rather speculatively from suppressed moralistic premises: what is moral is fitting for the divine, while nothing immoral is fitting. There is not anything obviously immoral or unfitting with regard to having a mind. By contrast, Parmenides's conception of divine nature is based upon ontological and modal considerations—and probably other generally held assumptions like the impossibility of creatio ex nihilo and Principle of Sufficient Reason—and he reasons deductively from those principles. Parmenides can thus be seen as following where inquiry into such natures led, what argumentation could be offered from the conception of necessary being itself, and abandoning the qualities that were not directly implied by such or inconsistent with that mode of being. Since thinking involves activity, and at least internal subjective changes, a perfectly complete, motionless, and unchanging entity cannot consistently be allowed to do so since this would imply some lack of eternal perfection.

Furthermore, while this is indeed a detraction from Xenophanes's conception of divinity, it does not appear to be all that novel in the historical context since conceiving of divine nature as lacking mental capacities is apparent in earlier Presocratic thought. Anaximenes's aēr is described as "divine," and though it is "active" in the world in some sense, there is no hint of mental intentionality to this fundamental "substance."[72] Similarly, Anaximander's conception of the apeiron as a "divine being" (one that is also eternal and indestructible), though it is said to "steer all things" (15a), seems to do so from natural necessity rather than mental intentionality. Given this, the fact that Parmenides's "God" doesn't *do* much of anything can be seen as simply taking the next logical step along the general path of the Presocratics, moving further away from the traditional religious conceptions of divine nature to a complete elimination of all anthropomorphic attributes and activities, including even mental activity.

Conclusion

The advantages of this divine-modal reading of Parmenides are numerous. Recognizing Doxa as a mythopoetical theogonical cosmogeny to be rejected, and Truth as a corrective on mortal views about divine (necessary) being, readily resolves the pervasive T-D Paradox. In turn, it also makes the Proem interpretively relevant, as it provides an introduction to the general subject matter and serves as a foil that will be later paralleled and rejected in Doxa. On these counts, this interpretative framework offers a far more holistic account than previous attempts have been able to offer. By further limiting the scope of Truth to divine (necessary) being, it still avoids the deeply problematic views of Parmenides as an extreme monist.

To avoid all of these problems, as well as the deep problem Palmer himself runs into by trying to make Doxa a positive account of contingent being, one need only accept that Parmenides was in fact closely following Xenophanes, as so much evidence supports, and that Parmenides simply took the next logical step offering metaphysical arguments for a "rational theism" as a corrective against traditional religious conceptions of divine nature rather than a metaphysical analysis of all being in-itself.

This reading greatly rewrites Parmenides's legacy. His novel metaphysical advancement is recognizing, and providing deductive-style argumentation for, the qualities required by any necessary mode of existence. In doing so, he also advanced inquiry into the philosophy of religion, identifying divine nature with that necessary mode of being, and thus raising new challenges to the mythopoetic traditions. It is quite likely this identification also came from Xenophanes's view that it is just as impious to say the gods come to be as to claim they can die, and perhaps even Epimenides's stark criticism of the Cretans's view that Zeus died. In the end, Parmenides can be understood as providing the most radical and absolute treatise from ancient Greece in which the divine is denied anthropomorphic qualities. This even suggests a substantive and nonmystical reason why Parmenides may have chosen to write in such ambiguous and enigmatic way: challenging deep cultural traditions, particularly religious views, can be a very sensitive and even dangerous activity.

Notes

1. Palmer 2009.

2. The initial development of the ideas and argumentation in this chapter can be found in parts of my doctoral dissertation, DeLong 2016, chapters 6–7.

3. Lewis 2009; Owen 1960, 94 n. 2; Wedin 2011.

4. See Colin Smith's chapter in this volume for discussion of the recent history of this passage's interpretation.

5. Concerning 2.3, Palmer (2009, 93–100) provides an overview of different possible senses of εἶναι, and presents good reasons for rejecting pure veridical and predicative senses and maintaining the standard existential meaning as the primary focus. However, he also acknowledges the "fused" (existential and predicative) nature of the Greek verb and suggests that either sense can be emphasized as necessary, while the existential sense remains primary for Parmenides's own argumentation, especially in Fragment 8. This particular translation from Palmer attempts to make both possible senses simultaneously clear.

6. This requires translating χρεών ἐστι at 2.5 as "must be," or "necessarily is," allowing an existential usage rather than a purely normative sense, for forms of χρή in Archaic Greek: a usage that some significant scholarship has otherwise denied for this time period. However, novel usage must enter the language at some point for it to have that usage later. Furthermore, it is quite difficult to make sense of the second route of inquiry if χρεών ἐστι is read with more traditional Archaic senses.

7. Palmer 2009, 103–5.

8. Rarely, and more controversially, Palmer (2009, 125–33) takes other forms of ἔστι besides the substantive participle also to serve as shorthand for the modal descriptions, as needed.

9. Frs. 2.6–8; 7.1–2; 8.7–9a; cf. Palmer 2009, 86–105. This reading makes far better sense than explaining this point from Russellian concerns with negative existential statements, for example. As Palmer argues, the association between

necessary inconceivability to impossibility is a far more likely inferential consideration to have been recognized in the historical context than the problem of negative existential statements and their lack of referents.

10. Line 2.4; cf. 6.6–9 for the contrast with the third route of inquiry as "wandering."

11. Fragment 3: . . . τὸ γὰρ αὐτὸ νοεῖν ἐστίν τε καὶ εἶναι. Michael Wiitala's chapter in this volume concerns the meaning of this fragment.

12. While I accept most of Palmer's views on Truth, I do not follow him in holding that the necessary being described therein must exist "everywhere," extended throughout the cosmos coextensively with contingent being.

13. Palmer 2009, 159; Palmer 2012.

14. A common interpretive stance based upon Fragments 2, 3, and 8.7–9a.

15. A common way of translating and understanding αἵπερ ὁδοὶ μοῦναι διζήσιός εἰσι νοῆσαι (2.2).

16. This is a highly contentious point, which I cannot develop and fully defend here. Let it be noted that the reasons for attributing a distinct third route of inquiry hold independently of accepting the modal interpretation. For an extended defense on this point within this collected volume, cf. Matthew Evans, "How Many Roads?"

17. The modal reading once again readily explains this epistemic inferiority, due to the third route lacking the a priori deductive certainty available on the initial two routes.

18. While there is some reason to worry about anachronistic Platonization in the epistemological distinctions since this distinction is relatively basic and not accompanied by a corresponding Platonic ontological hierarchy, the concern is not nearly as anachronistically troublesome as other "Two-World" views advanced by many other commentators. Some of the most influential examples of this reading include Nehamas 1981, Cosgrove 2014, and Thanassas 2007.

19. Palmer 2009, 106. Palmer emends the text of Fragments 5 and 6 to provide a far more positive spin on Doxa than would otherwise be present in the poem.

20. The most influential examples of this reading include Tarán 1965; Kirk et al. 1983; Coxon 2009; Nehamas 2002; Furth 1968; Guthrie 1965; Long 1975; Wedin 2011; and Owen 1960.

21. Palmer 2009, 114–18.

22. Palmer's view (2009, 163 and 180) also appears to be inconsistent on this point, by claiming mortals can "know" things about the contingent world in his treatment of cosmological claims.

23. Diels provided the otherwise missing, bolded verb [εἴργω, "I restrain"] at line three, based upon a parallel passage at Fragment 7.2 in which the youth is also commanded to "restrain thought" (εἶργε νόημα) from the path of mortals.

24. χρὴ τὸ λέγειν τε νοεῖν τ' ἐὸν ἔμμεναι· ἔστι γὰρ εἶναι, / μηδὲν δ' οὐκ ἔστιν· τά σ' ἐγὼ φράζεσθαι ἄνωγα./ πρώτης γάρ σ' ἀφ' ὁδοῦ ταύτης διζήσιος

<εἴργω> / αὐτὰρ ἔπειτ' ἀπὸ τῆς, ἣν δὴ βροτοὶ εἰδότες οὐδὲν. / πλάττονται, δίκρανοι. ἀμηχανίη γὰρ ἐν αὐτῶν / στήθεσιν ἰθύνει πλακτὸν νόον· οἱ δὲ φοροῦνται / κωφοὶ ὁμῶς τυφλοί τε, τεθηπότες, ἄκριτα φῦλα, / οἷς τὸ πέλειν τε καὶ οὐκ εἶναι ταὐτὸν νενόμισται / κοὐ ταὐτόν, πάντων δὲ παλίντροπός ἐστι κέλευθος.

25. Palmer (2009, 65–67 and 112–14) reads the opening description of Fragment 6 to be a further defense of the first route (i.e., necessary being) alone. However, the goddess could hardly be warning the youth away from that route, which is the recommended route in Truth. Thus, Palmer rejects Diels's emendation (εἴργω) at 6.3 and substitutes "I shall begin" (ἄρξω), resulting in: "I shall begin (ἄρξω) for you first from this (the route of necessary being), and then from this other route (mixed route), upon which mortals wander." Palmer's own proposed emendation is far from convincing, and is problematic in its own ways. For more extended argumentation against Palmer's reading, see Wedin 2011, 48n66 and Evans's chapter in this book.

26. The parallel from which Diels amended Fragment 6. Palmer (2009, 124) freely admits the goddess is indeed warning the youth away from the "mixed path" in Fragment 7.

27. τόν σοι ἐγὼ διάκοσμον ἐοικότα πάντα φαντίζω / ὡς οὐ μή ποτέ τίς σε βροτῶν γνώμῃ παρελάσσῃ. Palmer's translation reads πάντα as an adjective modifying διάκοσμον rather than ἐοικότα to capture the purported "exhaustiveness" of the goddess's cosmology. It is quite speculative to infer a far more "exhaustive" treatment than that evidenced by the extant fragments. There is no mention in the testimonia of any subject matter that is missing in the extant fragments, and Parmenides's verse is quite condensed.

28. Palmer 2009, 162n35.

29. Palmer (2009, 165–66) even admits Parmenides provides such an explicit warning against taking Doxa to contain knowledge that does not wander, in Ebert's restoration.

30. Palmer 2009, 163.

31. This latter point shows why the traditional focus of referring to Doxa as a "cosmology" is inaccurate and misleading at best. In fact, Doxa begins with a cosmogeny and theogony.

32. Even if one wants to argue that his source is the goddess, this merely puts Palmer's theogony at the very same level of justification as Hesiod's appeal to divine revelation via personal experience, for which there is no objective ground to determine either to be "superior."

33. Palmer 2009, 352–54.

34. Thus, Palmer's view cannot simply be saved by rejecting Ebert's restoration.

35. This would be uncontroversial in either placement for 8.34–41.

36. Parmenides's use of "mortal naming" should be taken to mean "specifying the nature of x," rather than merely assigning arbitrary reference signs, as Palmer (2009, 168–69) correctly notes.

37. Palmer 2009, 172. Palmer (2009, 167) further admits (implicitly, at least), that the Naming Error requires mortals getting something wrong about their conception of Truth itself—and not just that they are ignorant of that mode of being—when he notes that mortal naming is "meant to express the way mortals typically, though misguidedly, conceive of things."

38. Palmer 2009, 172.

39. Palmer 2009, 324–36.

40. Some representative skeptical examples are Coxon 2009, 18–20; Graham 2010, 4 and 95; McKirahan 2010, 151–52 and 18–20; Cordero 2004, 9–11; Curd 2004, 16n33; Patzia 2016; Tarán 1965, 3 and 201.

41. I have developed substantial and detailed arguments for this in DeLong 2017 and 2018a.

42. Cf. Aristotle, *Metaphysics* 986b20–23; Theophrastus, *Opinions of the Natural Philosophers* fr. 6 (Coxon *Test.* 40); Sextus Empiricus *Adv. Math.* vii, 111–14 (Coxon *Test.* 136); Aetius, *On Principles*, i, 3 (Coxon *Test.* 55); Pseudo-Plutarch, *Stromata* 5 (Coxon *Test.* 87).

43. Clement, *Misc.*, 5.109; fr. B23. Sextus, *Adv. Math.*, 7.49.110; fr. B34.

44. Sextus, *Adv. Math.*, 9.144; fr. B24.

45. Simplicius, *Comm. Physics*, 23.19, 23.10; fr. B25–26.

46. Inferred from Xenophanes's rejection of portrayals that make the gods as anthropomorphic beings, engaging in inappropriate behaviors like thieving, adultery, war, deception, and so forth; cf. fr. B1, B11–15.

47. Clement, *Misc.*, 5.109; fr. B14. Aristotle, *Rhetoric* 1399b6–9; A12.

48. Aetius (*On Principles* i, 3; Coxon *Test.* 55 and *Whether the Cosmos Is Indestructible* 11; Coxon, *Test.* 60) best describes this parallel, claiming both held the cosmos to be ungenerated, eternal, imperishable, unmoved, unique, and complete. Cf. Pseudo-Plutarch *Stromata* 5 (Coxon *Test.* 87). More moralistically, Cicero (*Lucullus* 129; Coxon *Test.* 102) claims both held the sole good to be that which is one, alike, and always the same.

49. Sextus, *Adv. Math*, 7.49.110; fr. B34. Cf. fr. *Xen.* A24. Cicero, *Lucullus* 74 (Coxon, *Test.* 101); Sextus, *Adv. Math.*, 111–14 (Coxon, *Test.* 136).

50. Fr. B11–12, B14–16. Stobaeus, *Selections*, 1.8.2; fr. B18.

51. Palmer 2009, 331.

52. Palmer 2009, 324–31.

53. Simplicius, *Comm. Physics*, 23.19; fr. B25.

54. Aetius, *On God*, 26 (Coxon *Test.* 56). A nearly contemporaneous (though less reputable) text similarly states that Parmenides was describing God, in largely Xenophanean terms, with one notable departure: Parmenides holds that God is limited ("like a sphere"), rather than unlimited; see Pseudo-Aristotle, *On Melissus, Xenophanes, and Gorgias*, 978b7–15 (Coxon *Test.* 120).

55. Philo of Alexandria, *On Providence*, ii, 39, 42 (Coxon *Test.* 104–5).

56. Cf. Pseudo-Aristotle, *On Melissus, Xenophanes, and Gorgias*, 978b7–15 (Coxon *Test.* 120). Menander Rhetor, *The Division of Epideictic Speeches*, I, 2 and

5 (Coxon *Test.* 151–52); Clement of Alexandria, *Stromata*, v, 14 (Coxon *Test.* 129–30); Ammonius, *Comm. De Interpreatione*, 133, 16–24; 136, 17–22; (Coxon *Test.* 187–88); Macrobius, *Comm. On Scipio's Dream*, I, 2, 20–21 (Coxon *Test.* 161); Boethius, *Consolation of Philosophy*, iii, 12, 96 (Coxon *Test.* 218).

57. The reliability of the sources is admittedly far from ideal, with anachronistic biases quite likely, especially from Neoplatonist and Christian perspectives. However, this evidence should not be summarily dismissed, for these earlier commentators likely had far superior access to complete texts, knowledge of other lost commentaries and sources, and much better understanding of the linguistic and cultural contexts. Thus, the default presumption should be to lend some credence to such reports unless good reason for skepticism can be provided for each.

58. More developed argumentation for the points in this section is in DeLong 2017. On how these parallels support understanding the poem as a ring-composition, see DeLong 2016, ch. 5.

59. Scholars debate whether the youth's journey indicates (a) anabasis: metaphorical ascent from darkness (ignorance) to light (knowledge), or (b) katabasis: a descent back to the House of Night. Though I favor the katabasis interpretation, this preference makes little difference here.

60. For this, see especially Mourelatos 2008, 1–46.

61. The sole source for lines 1.1–27 is Sextus Empiricus. Not only do no other ancient sources quote from these lines, no other source ever mentions their content. Sextus himself offered a clearly anachronistic, allegorical treatment of the Proem. Allegorical approaches are the second most common treatment. However, these are quite problematic in the details, and ultimately offer little-to-no interpretative impact.

62. The description of the chariot's "fire-blazing wheels" (1.6–8) also helps confirm this.

63. That the journey ends in Tartarus, having completed a full circular journey, is also supported by the "dark yawning chasm" beyond these gates (1.17).

64. Cf. Miller 2006.

65. As indicated by these programmatic framing fragments: 8.34–41, 8.51–61, 9, 19.

66. Likely drawing on similar mythopoetical sources; see especially Phaedrus's speech in Plato's *Symposium*.

67. While Fragment 18 is only now attested in Latin rather than Greek, the Latin translation seems to have picked up on this theme, and intentionally capitalized the reference to Aphrodite (Veneris, or Venus).

68. Cf. particularly Xenophanes B1, 10–12, 14–16.

69. Fragments 11, 12, 26. There is also a case to be made, I think, that Xenophanes's major work on this topic is arranged quite closely in structure and content as Parmenides's own.

70. For Xenophanes's description of his supreme God, see particularly fragments 23–26.

71. Cf. Palmer 2009, 329–30.

72. This divine substance also comes to be and is unlimited, unlike Parmenides's more Xenophanean conception. Nevertheless, nothing prevents Parmenides from combining these conceptions; cf. Xenophanes fr. A10.

Chapter 2

Parmenides's Poem as Initiation

Mary Cunningham

Parmenides frames his poem as the story of a youth's encounter with a goddess and his reception of her divine insight. There are many ways to understand the youth's experience: it is educational, religious (at least insofar as the youth is interfacing with a goddess and other daimones), and transformative. In my chapter, I aim to probe the philosophical meaning and implications of this dramatic framework by analyzing the youth's journey as an instance of initiation. This is in contrast to interpretations of the poem as religious (e.g., Orphic) or secular.[1] By foregrounding the initiatory structure of Parmenides's poem, a certain set of issues emerges as central. When we view the poem as an act of initiation, we ask ourselves: who is being initiated, and into what new status or group? What is the nature of the initiand's transformation? Why is initiation required or desirable for the initiand? I find that there is something to be gleaned from the ways Parmenides departs from and appropriates rather than adheres to the traditional model of initiation in his work. Consequently, I aim in my chapter to explore the aspects of Parmenides's poem that lie in the margins between tradition and what we would now recognize as philosophy.

The structure of the chapter is as follows. I begin with a reflection on what I take to be the crucial components of initiation in various contexts of Greek culture, taking direction from Walter Burkert's analysis of the structure of initiation rites.[2] I focus primarily on the Proem to identify the initiatory elements of the youth's journey, but also look to certain key sections from Truth. After characterizing the role of initiation in the poem, I present two ways that Parmenides's poem diverges from the traditional structure of initiation.[3] In these presentations, I examine the way that Parmenides appropriates certain elements of initiation, providing a critique while also expanding the poem's sense of initiation.

The Structure of Initiation

To situate the role of initiation in Parmenides's work properly, I will begin with a few remarks on the nature of initiation as such. Before I begin, however, I should make an important clarification. I am interested in the ways Parmenides's poem is construed as an initiation in a general sense rather than a specific one. I will be emphasizing the ways that Parmenides makes use of a general structure that is common to all initiation rites, not drawing parallels between Parmenides's poem and any particular historical initiation rites. As a result, I will be concerned here with identifying the structure that is common to all or most initiation rites.

As a practice, initiation does not belong to any one sphere of Greek culture; it is not exclusively political, domestic, or religious. Rather, it is a foundational structure that unites these different spheres of life due to a common care for rearing and educating youth. Insofar as the population of any polis is constantly replacing itself with a new generation, it is necessary for the polis perpetually to introduce the new generation to its various traditions, hierarchies, and structures that constitute political order to maintain the identity of the polis. Initiation functions as this introduction to and preservation of essential cultural structures for maintaining an intergenerational collective identity. Burkert suggests that its crucial public role as "the formation of the rising generation" makes initiation a fundamental practice for any civilization, evidenced by the well-known role of coming-of-age rituals in early cultures.[4] Initiation functions as a way of maintaining a certain established way of life in the face of human mortality.

We should also think of initiation as a practice for the individual. At its core, initiation is a kind of transformation of the individual involving some change in status. Sometimes this is based on human maturation, with adolescents being initiated into adulthood in coming-of-age ceremonies.

In Attica, for instance, there was the ritual death and rebirth of young girls into women at the sanctuary of Artemis at Brauron, and the cutting of the hair and introduction into the phratry of boys in the koureion.[5] In each of these cases, a child took specific measures to transform into an adult. In other contexts, initiation was not limited to the concerns associated with a change in bodily maturity, but with a transformation of the whole person.[6] Here, we might consider religious initiation rituals, like the mysteries conducted at Eleusis, which involve a personal transformation that has implications for one's state after death.[7] In these cases, the individual is transformed mentally or psychologically as a result of receiving some secret, enlightening knowledge, or because the completion of the rites allows the initiate some special protection from a god in the afterlife. Secrecy is often associated with initiation, especially in the religious sense, but this is not always the case. Some initiation is not only public, but a spectacle, as in the Spartan whipping ritual for the ephēboi at the sanctuary of Artemis Ortheia.[8]

The transformative quality of initiation is of a certain character, distinct from related phenomena like education or theōria.[9] To begin with, initiation is a personal transformation that takes place in the context of a designated, unique, and carefully curated setting. Burkert provides a tripartite analysis of the transformative nature of initiation rituals.[10] This model of initiation consists of (1) the isolation of the initiand, (2) the occurrence of some personal transformation in a liminal or foreign space, and, finally, (3) reintegration into regular life.

Insofar as initiation is transformative, it is inextricable from purification. Purification—which we will take up at length later in the chapter—creates the conditions for initiation, and initiation cannot be considered outside of the purificatory context. This is because purification in the Greek tradition constitutes entry into the liminal space in which the initiand's isolation takes place, and from which the initiand experiences transformation. In even the most mundane rites, purification serves to set a boundary; it establishes the limit and standard of cleanliness, that is, ritual fitness for initiation. In short, this model of initiation is inconceivable without purification as the step that demarcates the standards for an individual's journey into transformation.

Burkert's tripartite model of initiation maps the contours of the kouros's journey in Parmenides's poem.[11] The initiand goes out, is transformed, and then returns to his regular life. The first fragment opens by drawing attention to the youth's separation from everyday life: "The mares that carry me kept conveying me as far as ever my spirit reached" (1.1). Parmenides continues to emphasize the degree of separation from the youth's daily life, from his description of the speed and ferocity at which the chariot carries him away ("the axle in the naves kept blazing and uttering the pipe's loud note, driven onwards at both ends by its two metalled wheels," 1.4–5), to the words he has the goddess first speak to the youth ("Welcome, O youth, arriving at our dwelling as consort of immortal charioteers and mares which carry you; no ill fate sent you forth to travel on this way, which is far removed indeed from the step of men, but right and justice," 1.24–25). Parmenides takes care to let the reader know that the youth is far from home.

In the youth's journey, the second component of traditional initiation (i.e., the initiand's transformation) is first hinted at as the goddess receives the youth: "You must be informed of everything, both the unmoved heart of persuasive reality and the beliefs of mortals, which comprise no genuine conviction; nevertheless you shall learn these also" (1.28–31). At first, the youth's transformation is presented as a primarily intellectual process. The youth's journey, as the goddess sets it out, will be an educational one. However, as we will examine further below, Parmenides gives us reason to question the nature and scope of the youth's transformation.

Finally, the youth's journey completes the process of a traditional initiation in his return to daily life. While Parmenides does not give us a detailed picture of what this looks like for the youth, he has the goddess emphasize the importance of the youth's reintegration into daily life in light of his education. She begins her lesson with reference to what the youth should do with it once he has been educated: "Come now, I will tell you, and do you preserve my story, when you have heard it" (2.1). With these three basic components—the journey away from daily life, personal transformation, and reintegration into daily life—Parmenides's poem tracks a journey that is recognizable as initiation. However, while the poem follows the basic structure of an initiation, the kouros's journey also differs from the usual model in significant ways. Rather than continuing to outline the ways that Parmenides's poem resembles an initiation ritual, I'll offer a series of reflections on the areas in which the poem appropriates aspects of traditional initiation. I will focus on two general ways that Parmenides appropriates elements of initiation in the Proem and Truth. First, considering Fragments 1 and 2, I will address the provocative absence of a purification accompanying the kouros's initiation. Next, focusing on the end of Fragment 8, I will reflect on the indefiniteness of the kouros's initiation, both in terms of its incompleteness and that the ritual is not clearly an enlightenment for the initiand.

Two Parmenidean Appropriations of Initiation

Initiation without Purification

Given its role in secluding the initiand from everyday life, purification is a crucial part of the first step of initiation. Purification functions as a kind of ritual drawing of boundaries, setting the initiand apart from the mundane or from pollutions.[12] Through various kinds of cleansing, purgation, or simply mental preparedness, purification delimits the initiand's interstitial status and establishes the distinctive space within which the initiand's transformation takes place. This essential condition for initiation is noticeably absent from Parmenides's account of the kouros's journey. The youth enters the purificatory space, separating from daily life at the gateway, without any purificatory action. The mares and maidens simply take him to the goddess as is.

As we have seen in Burkert's analysis of purification, this cannot be a mere omission. Given the lengths Parmenides takes to establish the religious and ritual setting of his poem, it is difficult—if not impossible—to accept a lack of purification as anything but a statement in its own right. There is certainly an initiation taking place, but from what and toward what? Without the accompanying purificatory rite, there is no standard set between the fit initiand and the profane world around him, and therefore no context for the kouros's initiation, what he is becoming, or what he is learning. Given the inseparability of purification and initiation, and given the clear indication that this is an initiation, we must inquire into the standard of initiation for ourselves.

Parmenides's poem both emphasizes and calls into question the distinction between "is" and "is not," and the loud omission of purification seems to me to amplify this double inquiry. What, indeed, is purification without an initiation? Why, or *are*, these mutual opposites necessary, and how do they constitute the grounds for individual growth? One might wonder whether this relates to the way Parmenides's poem seems to perform a reversal of the usual relationship between initiation and the history of the practice of documenting such rites. While initiation rites are not always secret affairs, they often are.[13] In the case of cult initiation, for instance, the details of the rites are kept secret and usually not disclosed in writing.[14] The artwork and literature associated with the cult deal with subject matter that is public knowledge, such as the stories and imagery of the cult's human founders, the names and characteristics of the gods worshiped, and the origins of the rites. This is the opposite of what Parmenides reveals in his poem, both dramatically and philosophically. At the level of dramatic framework, the poem discloses the initiation rites directly, revealing what is normally kept secret and omitting what is usually the topic of the literature associated with an initiation (i.e., the situating details). Parmenides does not identify the goddess as any specific familiar divine figure, calling her merely "goddess" (θέα, 1.22). In fact, Parmenides seems intentionally ambiguous about the kind of initiation he is presenting, providing reason to interpret the initiation's context in different ways. There are themes of both coming-of-age and religious initiation in his work. The initiand is specifically referred to as a youth (1.24), suggesting initiation in the context of maturity. At the same time, the youth is carried by daemonic figures (cf. 1.1, 1.9), and a goddess acts as his mystagogue (1.24), giving the ritual a distinctly religious tone. Parmenides takes special measures to de-contextualize the initiation that takes place so that the reader cannot identify it as one particular kind of ritual.

The dramatic reversal of the usual presentation of initiation rites mirrors Parmenides's treatment of the "is" in the poem (esp. 2.3). The "is," which is typically not the point of reflection, is brought into the foreground. Rather than have us focus on any particular subject or predicate, the goddess brings our attention to the "is" that in other usual circumstances underlies, connects, or makes room for the topic of reflection. Like the dramatic framing, this kind of inquiry is jarring, abstract, and unclear due to its lack of context. By necessity, there cannot be context for the object of Parmenides's inquiry, since what purely *is* is, by nature, not restricted to a limited context.

The issue of the poem's omission of a purification rite is complicated by the characterization of the youth. The way Parmenides introduces the youth suggests that he is not in need of purification prior to the rites. As I note above, Parmenides signals that the initiation taking place could be an induction into adulthood or special religious status. However, the identification of the initiand as "a knowing man" (εἰδότα φῶτα, 1.3) suggests that he does not need any special preparation for what is to come.[15] He is already prepared for enlightening knowledge insofar as he is knowing (εἰδότα) and for a coming-of-age ritual insofar as he is a man (φώς) as opposed to a child. In addition to the suggestion that the initiand does not stand in need of special preparation according to this initial characterization, Parmenides seems to emphasize a certain continuity in the experience of the kouros that stands at odds with the ritual boundaries required for purification.

The poem begins by establishing a continuity of experience between the kouros's daily life and his encounter with the goddess. The mares pulling the youth's chariot take him to the limits of his spirit (ὅσον τ' ἐπὶ θυμὸς ἱκάνοι πέμπον, 1.1–2), but not beyond that limit. The journey is situated specifically within the limits of the youth's capacities, rather than suggesting a broadening of or departure from his limits. However, this continuity in the youth's experience stands in stark contrast to the boundary presented by the gateway image. Through the interpretive lens of the youth's journey as an initiation rite, the Proem's gateway functions as a boundary marker at the entrance of the sacred space. The goddess, and her divine insight, lies beyond the gateway. When the goddess welcomes the youth, she calls the space beyond the gateway her home or dwelling (1.24). The gateway functions as a horos or a Herm, marking off the entrance to the sacred space or the space that belongs to the deity.[16] It should surprise us that the youth is able to cross this boundary without any special purification or preparation.

It is difficult to overstate the relationship between purification and boundaries in Greek religion. Purification functions as a boundary marker, which is especially necessary as a condition for physically entering a sacred space. This is true even in sacred spaces that are not as formally ritualized as cult spaces; the entrance to any temenos is marked with a lustral fountain for handwashing. With cult religion as one of the lenses through which we might understand the significance of initiation in the poem, an appropriate historical comparison to the gateway image is the Lesser Propylaea at Eleusis.[17] The Lesser Propylaea is the gateway that marks the boundary between Demeter's sanctuary at Eleusis and the Sacred Way, a road between Eleusis and Athens. In particular, the Lesser Propylaea marks off the entrance to the Telesterion, the covered structure within which the initiation rites are conducted. This gateway conceals and mediates access to the initiation rites at Eleusis both visually and ritually. The interior of the Telesterion is not visible from outside the gateway, and only purified initiands and initiates may pass through the gates.

Given the seriousness with which this gateway was treated in the context of the Eleusinian rites, one would expect the stakes to be even higher at Parmenides's gateway. This expectation makes the goddess's reception of the youth especially striking. The goddess greets the youth with a warm hello (χαῖρε, 1.26), and comforts him, telling him that no evil fate has brought him to the road on which he travels (οὔ τί σε μοῖρα κακὴ προὔπεμπε, 1.26). This kind reception seems to indicate that there is no expectation of a purificatory right that should accompany the youth's passageway through the gateway, which is a significant departure from a typical initiation.

The goddess's greeting complicates matters further. Even after crossing the boundary of the gateway, the goddess suggests a sense of continuity with the road behind the gateway. She tells the youth that the road (ὁδόν, 1.27) on which he travels is "far from the well-trodden path of humans" (ἀπ' ἀνθρώπων ἐκτὸς πάτου). Regardless of whether we understand ἐκτός as "far from" or "outside," the use of πάτος, the well-trodden or beaten path, rather than the simple ὁδός (road), communicates that while the youth has arrived somewhere far away, the place is not necessarily bound off from the path of humans. There seems to be a distinction between proximity and separation at play. The goddess could have chosen to supply ὁδοῦ rather than πάτου at line 27, which would have clearly set the youth's path entirely apart from the road of humans, delineating a distinctive initiatory space. Instead, she says that the road on which the youth travels is one that is far from the *well-trodden* path of humans, leaving room for us to understand that the youth is on a different part of the same road (i.e., the less-traveled part of the road).

Parmenides is provocatively ambiguous in his language and imagery. On the one hand, he presents the arrival at and passage through the gateway as a momentous occasion. The chariot shrieks as it approaches, the sun maidens throw the veils from their heads, and the gateway is described with care for the details of the way in which it opens, who holds the keys, and how one might pass through. However, for all the fanfare indicating the significance of the gateway as a boundary, the youth is ultimately able to pass through it as he would any other boundary, without some kind of accompanying purification to mark the sacred significance of the boundary. This care for boundaries in the poem's imagery combined with a seeming lack of ritual respect for those boundaries has to be considered an appropriation of this aspect of initiation on Parmenides's part.

I suggest that this appropriation serves as a critique of the concept of boundary, marking off, or setting apart in general. Boundary marking between opposite categories is revealed to be a faulty artifact of human thinking. The goddess proclaims that setting one's mind on absent things "will not sever Being from cleaving to Being" (οὐ γάρ ἀποτμήξει τὸ ἐὸν τοῦ ἐόντος ἔχεσθαι, 4.2), and that the way of thinking that names two forms (8.52), setting opposites apart from one another is "likely in its entirety" (ἐοικότα πάντα, 8.60). In light of this insight, we might think of purification as the ritual enactment of the faulty mortal tendency to name two forms set apart as opposites. Conceptually, purification represents a commitment to the distinction between the sacred and the mundane. Purification entails that, with respect to religious practice, there are two distinct and opposite categories or contexts of a person's condition—purity and pollution—and only one of these is conducive to religious experience that is appropriate, pious, or successful. In Fragment 8, Parmenides has the goddess say that this kind of boundary-setting is a misguided form of mortal thinking.[18] Consequently, to mark boundaries with an act of purification would be an inappropriate practice not befitting of the goddess's wisdom. Instead, Parmenides intimates both toward the distinction between sacred and profane and toward transcending this distinction. He gives the reader the appearance of such a boundary—the image of a physical, sensible gateway with swinging doors—but ignores the value of the boundary in ritual practice.[19] Hence, there is no purification rite for the kouros to undertake, nor is there an expectation that he would complete such a rite as a condition of his visit with the goddess.

The Indefiniteness of the Initiation

Shifting our focus from the outset of the youth's journey to its end, let us now consider the indefiniteness of the initiation. As an event of defined and total transformation, initiation is a process with a beginning and an end. At the end of the initiation, the initiate finds herself characterized by some new status. Aside from exceptional circumstances, this status is permanent and irreversible. Consider, for instance, the transformation that takes place in the case of a coming-of-age ritual. Once one has been initiated into adulthood, there is no risk that she will fall back into childhood, and there is therefore no need to renew one's adult status with further ritual at any point.

The kouros's initiation is different. To begin with, the nature of his transformation is not made clear. There is not a straightforward indication of how the kouros has been affected by the divine encounter, or how he has transformed compared to the outset of the poem. *That* the youth has been affected is clear, but *how* he has been affected is not.

The youth is described as being "knowing" (εἰδότα) at the beginning of the poem, making it unexpected that he would change as the result of gaining some new knowledge.[20] Further, there does not appear to be any kind of permanence or definition to the kind of transformation the youth undergoes as a result of his encounter with the goddess. There are still steps that he needs to take with respect to his initiatory transformation after he leaves the company of the goddess. At the end of Fragment 8, the goddess charges him with learning mortal opinions.[21] As the Doxa begins and the youth is re-integrated into daily life, he needs to sort out the deceiving order of the goddess's words for himself and still needs to hear about mortal opinions.

The transformative nature of initiation is meant to be definite insofar as it is edifying in a certain respect. The initiate leaves her daily life, experiences a transformation, and returns to daily life renewed. The edification of the transformation that defines initiation can be construed in several ways. In coming-of-age initiation rituals, edification takes the shape of maturity when one becomes an adult rather than a child. One's adult status permits a new, more defined sense of self. Alternatively, in the case of religious initiation, edification takes the form of a kind of spiritual or intellectual enlightenment. The initiate learns some secret rite or knowledge, which, in turn, influences her life. Understanding the general shape of the youth's journey as a philosophical movement out of ignorance and into knowledge, many have argued that our understanding of the philosophical insights of the poem depend on this interpretation of the youth's journey.[22]

There is good reason to understand the youth's journey as both a journey upward and a journey downward. To begin with, the mention of sun maidens (ἡλιάδες κοῦραι, 1.9) brings to mind their role in the myth of Phaethon, carrying his chariot up close to the sun.[23] Further, mention of the sun maidens leaving the house of night for the light (προλιποῦσαι δώματα νυκτὸς ἐς φάος, 1.8–9), and the description of the gateway being fitted closely to an architrave in the aither (αἰθέριαι, 1.13) also suggest an aerial orientation. On the other hand, that the goddess greets the youth by telling him that he has arrived in her company as the result of no evil fate (1.26) suggests that there is reason to expect that death should have led him to her presence. This view is underscored by the shrieking of the chariot's axle leading up to the passage through the gateway, an image evoking the ololygē immediately preceding the sacrifice.[24] Further, the House of Night, to which the gateway is the entrance, is located in Tartarus in Hesiod's *Theogony* (Hes., *Th.*, 744).

There is, however, a certain philosophical discomfort created by the katabasis interpretation. If darkness is meant to represent ignorance as is suggested at line 8.59 ("unknowing night," νύκτ' ἀδαῆ), then this view could imply the unexpected understanding of the poem's journey as a descent into ignorance rather than an ascent to enlightenment.[25] The ambiguity surrounding the direction of the youth's journey demonstrates one way that Parmenides seems to be intentionally de-contextualizing the journey, forcing the reader to focus on the insights of the poem alone without any orienting pretext or mythological context in which to rest imagination.

If initiation is a practice that traditionally edifies the knowledge or status of the initiate, then the youth's journey is untraditional insofar as it lacks traditional definition or edification. We need not be hard proponents of the katabasis interpretation to see a major departure from the traditional model of initiation here. As I note above, initiation as a practice aims to bring people into certain structures from without to uphold those structures. In other words, initiation is supposed to be an edifying practice. While different initiations have been modeled after various mythic ascending and descending journeys (e.g., rites established commemorating Herakles's apotheosis or his descent into Hades), they also must be knowledge-giving or edifying insofar as they are transformative. Because of this, any degree of ambiguity or uncertainty about whether the initiate receives knowledge, or the content of that knowledge, is a radical divergence from the typical initiation. Whether Parmenides's initiate receives edifying knowledge or status is unclear.

The goddess's parting words just prior to the start of Doxa suggest another way that the youth's initiation is indefinite. Unlike a typical initiation, the youth's initiation appears to be incomplete when he leaves the company of the goddess. After the goddess finishes her trustworthy speech (πιστὸν λόγον, 8.50), she charges the youth to "learn mortal beliefs, hearing the deceptive order of [her] verse," (δόξας δ' ἀπὸ τοῦδε βροτείας μάνθανε κόσμον ἐμῶν ἐπέων ἀπατηλόν ἀκούων, 8.51–52). The goddess goes on to explain that the mortal opinions the kouros should learn are designations of opposites—two forms set apart from one another—an example of which is the distinction between light and darkness (8.56–59). She characterizes mortal opinion about opposites as the belief that each opposite is "the same with itself in every direction but not the same as the other" (8.57–58, trans. Coxon) and calls this way of thinking entirely *likely* (ἐοικότα πάντα, 8.60), that is, falling short of the truth.

This task the goddess charges the youth with is strange in the context of an initiation for two reasons. First, the initiation ritual appears to be incomplete when the youth departs the sacred space and the company of the goddess. As opposed to a historical initiation ritual, which takes place within defined limits of space and time, the youth's initiation extends back beyond the defined boundary of the gateway. His initiation into her divine thinking requires ongoing education and renewal rather than a single event. This is to say, learning the goddess's wisdom is not enough; the initiate must still go forward and learn mortal opinions in addition to the divine knowledge. Further, the nature of what he needs to learn when he leaves the goddess adds to the strange indefiniteness of his initiation. With initiation being an essentially edifying practice, transforming the individual by means of some special knowledge or insight, it is wholly unexpected that mortal opinions, in which there is no true conviction (1.30), would play a role in the youth's experience.

Here, I find it helpful to return to the idea of boundaries. While drawing boundaries is the function of purification, initiation too relies on a certain set of boundaries between opposites. Without the distinction between sacred and profane, knowledge and ignorance, or adulthood and adolescence, there is no meaningful function for initiation as a practice. As discussed above, initiation is a practice that fundamentally transforms the individual. That is to say, to be initiated is to become a new kind of person in some respect (e.g., with respect to one's maturity, religious status, civic status, etc.) To take initiation seriously is to see the world in terms of normative pairs of opposites of which one is either enlightened as an initiate, or otherwise ignorant, without a meaningful continuum between those two states. In this way, initiation is a cultural practice that orients its participants to the world by way of the very framework of opposition that the goddess calls "wholly likely" in Fragment 8. In light of the goddess's words about this framework, we might consider how this relationship between initiation and conceptual opposition invites us to consider the ways in which the kouros's journey is not constrained by the likely boundaries of mortal opinion. The poem invites us to consider what is basic to the pairs of opposites that themselves seem basic to everything (i.e., hot and cold, low and high, light and dark, etc.).[26]

Concluding Remarks

In this chapter, I have outlined some of the many ways that Parmenides's poem both conforms to and departs from the structure of initiation in the ancient Greek world. By isolating the character of initiation as both a cultural and a subjective practice, I have identified how the poem's framing relates to, and stands at odds with, some of the poem's philosophical content.

There is a fundamental tension between the poem's organization and its key insights. If Being is common to all, why does the youth require a special initiation rite? Why would Parmenides choose the esoteric framework of initiation as the organizing theme of his poem, while at the same time having the goddess attack the notion of dichotomy as entailing independence and mutual exclusivity? I think we cannot help but see this as an intentional provocation from Parmenides. If the various significant interpretive ambiguities in the poem—such as the direction of the chariot's journey—are meant to invite us to consider the insights of each interpretation as being held in common insofar as the poem is aimed at revealing what is in common (i.e., Being, or what is), then it seems right that something similar would be occurring at the level of the form of the poem in general.[27] Parmenides is careful in his appropriation of initiation. He uses enough of core features to make it identifiable as something informing the youth's journey in the poem, while at the same time making the journey a reversal of many of the key features of initiation. In this way, Parmenides invites us to consider, as we have in this chapter, what is common between an initiation and the youth's journey. Alternatively, we may consider the poem a critique and reversal of initiation since, as we have discussed, Parmenides seems to question the philosophical meaningfulness of transformation within the context of oppositional boundaries by critiquing a certain conception of opposites as entailing mutual exclusivity.

Notes

1. Many scholars have analyzed the cultural significance of the poem's framework. For instance, Bowra's analysis of the Proem emphasizes the religious overtones of the youth's journey, drawing attention to resemblances between certain aspects of the Proem and elements of Orphism (cf. Bowra 1937, 109). Cosgrove

(2011, 30) later interpreted the journey as a more secular philosophical endeavor, arguing that the journey stands in contrast to the popular conception of natural philosophy at Parmenides's time. I view my chapter as operating within this tradition of interpreting the intersection of philosophy and religion in Parmenides's work.

2. I rely on Burkert's analysis of initiation from his text *Greek Religion* (1985), especially 543–52, "Initiation."

3. My project is indebted to Mitchell Miller's work on Parmenides, especially Miller 1999 and 2006. The former has contributed to my understanding of the initiatory structure of the poem (although Miller does not characterize the structure of the poem in these terms). Further, the latter has been crucial for my understanding of the Proem's imagery—especially the significance of the gateway image—and the ways its ambiguity unlocks the most philosophically rich elements of the poem. I am especially indebted to the methodology of Miller 1999, in which Miller observes some of the ways that Plato's work, which in many ways resembles Greek drama, appropriates certain dramatic elements to address some of the problems with mimetic drama discussed in the *Republic*. I have adopted this general approach in my project.

4. Burkert 1985, 544. See Eliade 1958 for a more complete exploration of this idea.

5. See Cole 1984 for a detailed analysis of the social function of each of these initiation rites.

6. I want to stress here that there does not appear to be a rigorous or even functional distinction between soul and body that delineates or informs the consequences of different kinds of initiation. In fact, this distinction is often misleading to make in these kinds of conversations.

7. See Burkert 1987 and 2012, esp. 595–605 and Mylonas 1961, esp. 224–86 for more on the nature of the personal transformation allowed by the initiation rites at Eleusis.

8. Burkert 2012, 548.

9. See Nightingale 2004 for the nature and philosophical function of theōria, and in particular 32–34 for observations about Parmenides's poem with reference to this cultural phenomenon.

10. He writes, "The distinctive mark of initiation is the temporary seclusion of initiands from everyday life to a marginal existence. The ritual proceeds through three stages of separation, interstitial status, and re-integration" (Burkert 1987, 544).

11. Other scholars have recognized this structure of Parmenides's work. Miller (1999) characterizes the structure of Parmenides's poem as following these four parts: (1) elicitation of highest human insight, (2) disclosure of the limits of that insight, (3) disclosure of divine truth, and (4) return to human opinion. Miller attributes the identification of this general pattern in Plato's dialogues to Ketchum (1981). However, the observation that Plato seems to take this structure over from Parmenides's poem, along with the mapping of the four-part pattern onto Parmenides's poem, is Miller's.

12. For the relationship between purification and physical or ritual boundaries, see Burkert 2012, esp. 162–65, Parker 1996, and Horster 2010. The Eleatic Stranger's reflections on this relationship in Plato's *Sophist* (226e–227d) are useful as well.

13. See Burkert 2012, 576–634 for many examples of secret initiations in mystery religion.

14. In some cases, decreed by law, as the secrets of Eleusis were protected by Athenian law (cf. Mylonas 1961, 224).

15. On the difficulty surrounding the interpretation of εἰδότα φῶτα at 1.3, see Cosgrove 2011.

16. See Horster 2010 for more on the gods' ownership of physical spaces and the role of physical spaces in Greek cult.

17. For more on the details of the Sanctuary at Eleusis, see Mylonas 1961 and Evans 2002.

18. On this issue, Cherubin 2004 and Miller 2006 are particularly helpful.

19. Cf. Miller 2006, 12–18 on the significance of the gateway image and the issue of opposites.

20. I owe the identification of the strangeness of this characterization of the youth to Cosgrove 2011. While, as Cosgrove notes, the standard interpretation of this line understands εἰδότα φῶτα as referring to the youth (or, as Cosgrove calls him, the "Parmenidean initiate"), Cosgrove argues that the "knowing man" is a different character entirely. I will be operating with the standard interpretation that the youth is the one picked out at 1.3.

21. ". . . from this point, learn human beliefs" (8.51).

22. See Owens 1979 for a comprehensive summary and history of the key players in this debate and an assessment of its stakes. According to Owens's account, the direction of the youth's journey has been a matter of interpretive significance. Until the twentieth century, the prevailing view was that the youth makes an ascending journey from the depths of ignorance to the heights of knowledge, and, spatially, from the Earth's surface into the sky. Burkert (1969), following Morrison (1955) and Diels (1897), suggests that the youth undertakes a katabasis, a journey downward, rather than an Auffahrt, an ascent. Auffahrt proponents understand the poem as edifying and initiatory in the traditional sense. In fact, some see Parmenides, rather than Plato, as foundational for the allegorical journey out of darkness into knowledge and enlightenment. Nightingale (2004, 33–34) and Bowra (1937, 97–112) were both convinced by this view.

23. Bowra (1937, 103) and Owens (1979, 17), along with many others, draw this connection.

24. The high pitch of the ololygē is a distinctively feminine contribution to sacrifice and hence fits nicely with the overall feminine character of the figures who transport and instruct the youth.

25. In addition to these, there are views on this issue that transcend the debate between the Auffahrt and katabasis interpretations. Kingsley (1999) bridges

the gap, seeing the poem as an imitation of the Asclepian cult practice of incubation. On his view, the journey is a katabasis insofar as it mirrors a journey to a dark, cave-like dwelling, but still a ritual Auffahrt insofar as it is a practice of gaining special insight form the god. Miller (2006) argues that Parmenides is intentionally ambiguous about the direction of the journey, and I am inclined to agree with him.

26. Cf. Miller 2006, 28–29.
27. See Miller 2006, 12–18.

Chapter 3

Olympus as Hades

Plato and the Homeric Parmenides

ALEX PRIOU

Parmenides the Poet

The acclaimed discoverer of being was also a poet. He gave his speeches, the barest imaginable, a deliberately epic tone in both meter and motif.[1] Perhaps he meant his hexameters to announce the arrival of a new hero who has ascended a new Olympus, atop which resides a new god with a new law, an anonymous goddess who declares that being is one and that any division in it would require that nonbeing be. The continued discussion of this poem to this very day attests to what Parmenides had achieved. And though these verses and their hero never earned the admiration of the rhapsodes and their audiences, an honor enjoyed by their author's epic forebears, Homer and Hesiod, still they have enjoyed the audience of the most celebrated philosophic minds across millennia. This new hero's journey into the abstract, alien as it may be to most, remains very much our own.

This cursory overview of the poetic aspect of Parmenides's thought suits those captivated with the question of being, in that it ennobles the flight upward as a journey fit for a few and, like all rare things, of unparalleled value. Venturing beyond the bounds of our language and conventions, this new hero receives from a goddess the revelation about the defectiveness of human speech, which in mixing being with nonbeing speaks the unspeakable, thinks the unthinkable, names the unnamable, divides the indivisible, and violates the inviolable. Her prohibitions, however, have done little to settle the issue; they have rather, as Eva Brann remarks, "done for this philosophical offense what inveighings against sin have so often accomplished in the moral sphere—they have launched it on its career as a well-formulated and ever attractive presence."[2] Perhaps the goddess's shifts in tone, from the gentle to the punitive, presaged the legacy of controversy that her revelation eventually came to have almost immediately.[3] If the hero of this poem is our own, it is through leading us to the unity of being, only for it to serve as the field of battle wherein we have long done combat. His journey, like that of Odysseus, is one we happily learn about through hearsay, though so few of us are willing to follow in his footsteps.

Does our strained relationship with this hero, mixed as it is between admiration and criticism, mean that Parmenides's poem was a failure, that the goddess's revelation has fallen on friendly but skeptical ears? Or was it rather part of his intention to provoke such a quarrel? To answer these questions, I maintain, we must assess the influence of the epic tradition on the manner in which Parmenides presented his thought. That Parmenides borrowed from Homer is widely acknowledged, yet whether this amounts to a veneer or runs deep to the core of his intention is rarely asked, let alone understood. Studied primarily in light of other so-called Presocratic philosophers, Parmenides's relationship to the epic tradition fades from view. But how are we to approach this question when so little remains of his work? Perhaps the best guide in starting our journey is Plato, who gave his critique of Parmenides a Homeric cast. That is, Plato seems to have understood and expanded upon Parmenides's Homeric themes. Plato, then, might well shed light on this dark question. Toward this end, we will take a long detour through Plato's *Sophist* and *Statesman*, in which he advances his Homeric critique of Parmenides, as well as Homer's *Odyssey*, from which Plato borrows generously, before returning to Parmenides.

Socrates and the Cyclops

When, at the beginning of Plato's *Sophist*, Theodorus and Socrates meet up to resume their conversation from the day before, Theodorus surprises Socrates with an anonymous Stranger from Parmenides's home city of Elea.

From Theodorus's assertion that the Stranger is a very philosophic man and from his correction of Socrates's conception of philosophy, it is clear that Theodorus has brought the Stranger along to set Socrates straight. And who can blame him? In their conversation the day prior, as recorded in the *Theaetetus*, Socrates continually bullied Theodorus into joining the conversation, that he might take part in, even lead, the refutation of his late friend Protagoras. After having spurred his best student, Theaetetus, to chastise his elders, Socrates finally let Theodorus go; he remained silent for the rest of the conversation. Now, however, with the Stranger by his side, Theodorus freely chastises Socrates by asserting that the eristics of refutation have no place in philosophy. He evidently expects less contentiousness from the Stranger. The Stranger fulfills this expectation first by employing a method of dieresis far gentler than the Socratic refutation that so irritated Theodorus and, second, by maintaining that the sophist and the philosopher can ultimately be distinguished from one another, albeit with some difficulty. The Stranger maintains this over and against Socrates's own contention, which he evidently bases on the model of his own life, that the philosopher necessarily is confused with the sophist (and statesman) before the city. The Stranger proves to be a friendly visitor, bearing gifts for his host Theodorus.

Socrates immediately senses Theodorus's aim, comically exaggerating the gravity of the Stranger's arrival by suggesting that he may be, in accordance with Homer's speech, a god in human disguise.[4] Socrates alludes to two passages from Homer's *Odyssey*, both of which refer to Zeus's law of guest-friendship (ξενία).[5] The first passage is from Book 9, which finds Odysseus venturing into the cave of the Cyclops Polyphemus. Odysseus tells his audience, the Phaeacians, that he went among the Cyclopes to learn what sort of men they are, whether they be arrogant and savage, in no way just "or friendly to strangers, and their mind (νόος) is god-fearing" (*Od.* 9.175–76).[6] Based on Polyphemus's isolation from the other Cyclopes—they dwell high on the peaks of the mountains while he appears to dwell closer to the shore—Odysseus infers that he must have come to know greater lawlessness than the others.[7] He already suspects that the man he will encounter will be "garbed with great strength, savage, knowing well neither justice nor laws" (*Od.* 9.214–15). Upon entering the Cyclops's cave, however, he sees a level of organization and technical sophistication that suggests a man more civilized, hoping by staying (against the protests of his men) that Polyphemus will grant him what accords with Zeus's law concerning strangers (*Od.* 9.216–30). Polyphemus's first address to Odysseus and his men suggests that he may indeed adhere to Zeus's law; accordingly, Odysseus announces that he and his men are his suppliants and that "Zeus is the avenger on behalf of suppliants and strangers, the god of guests, who accompanies reverent strangers" (*Od.* 9.252–55, 269–71). Polyphemus, however, quickly dashes Odysseus's expectations, along with the brains of two of his companions.

With this first allusion, then, Socrates appears to make a joke about Theodorus's sensitivity to embarrassment in philosophic conversation. This sensitivity, Socrates suggests, has led him to construe a bit of vigorous argument, from the day before in the *Theaetetus*, for sophistry, in effect accusing Socrates of the discursive corollary of Polyphemus's cannibalism or more moderately, if we follow the second allusion, Antinous's violent abuse (as I discuss below). Theodorus did, the day before, compare Socrates to Sciron and Antaeus (*Theaet.* 169a9–b4). The Stranger, in turn, picks up on Socrates's allusion to Polyphemus, when he openly worries that "not to gratify you and these here, especially when you spoke as you did, appears to me something inhospitable (ἄξενόν) and savage" (*Soph.* 217e5–218a1). The Stranger presents himself as a potential Polyphemus but restrains himself from giving a long speech, instead conversing with another. The Stranger's restraint forces him to tailor his discourse to Theaetetus, who must understand and assent to the Stranger's proposals for the conversation to move forward. That is, the shift from long, continuous monologue to broken-up dialogue is necessary in order to be hospitable to one's audience.[8] The result of the Stranger's accommodation, however, is that Theaetetus will have difficulty in sighting the sophist, which will in turn require the Stranger to undertake an analysis of the sophist's use of spoken images (εἴδωλα λεγόμενα). But to account for such images requires a second, savage act: the murder of his intellectual father, Parmenides. For to account for images, in their both being and not being the thing imaged, requires violating the law of Parmenides's goddess. But if Parmenides blocks the way to true images of being, then he, and not the Stranger, seems guilty of savagery. Is Parmenides supposed to be Polyphemus, here?

Mind and Image

A glance back at Homer seems to confirm this suspicion. After Odysseus appeals to Zeus's law of ξενία (i.e., the relationship between host and guest), Polyphemus laughs, revealing that the Cyclopes have no fear of Zeus. Odysseus is confronted with a choice either to indulge his anger and kill his savage enemy now or to plot his revenge in prudence. Odysseus chooses the latter, devising to blind Polyphemus while naming himself Οὖτις, "Nobody." The sense of this plan becomes clear when, upon being blinded, Polyphemus cries out to the other Cyclopes, who awake to the sound and inquire as to Polyphemus's well-being:

> "Why, Polyphemus, do you shout like this in such distress
> through the immortal night and set us sleepless?
> Surely no one (μή τίς) among mortals drives your sheep
>   against your will?
> Surely no one (μή τίς) kills you yourself by deceit or by
>   force?"
> These, in turn, did strong Polyphemus address from his
>   cave,
> "Friends, Nobody (Οὖτίς) kills me by deceit, not by force!"
> And they, in replying, spoke winged words,
> "Well, then, if no one (μή τίς) does you violence, but you're
>   in such a state,
> then surely one cannot avoid a sickness from great Zeus,
> but *you* might pray to your father, lord Poseidon."
> They spoke thus, as they departed, and my dear heart
>   laughed,
> since my name deceived, and my blameless mind (μῆτις).
> (*Od.* 9.403–14)

To escape the Cyclops's cave, Odysseus tells him that his name is Οὖτις, Nobody. But since the other Cyclopes don't see anybody, they take Οὖτις as equivalent to μή τις, then depart before Polyphemus can clarify. Underlying Odysseus's trick is the observation that, because speech mediates our grasp of reality, it can be separated from reality and used to distort it. This is to present a being as what it is not, to craft the "spoken images" that make sophistry possible. This deception of names or words, ὀνόματα, leads Odysseus, via a pun, to identify the somebody who is nobody as something like the pure mind, μή τις as μῆτις.[9]

Like Polyphemus, Parmenides seems blind to sophistry. When the Stranger finally confronts Parmenides, he argues that, if being is one, as Parmenides claims, then these two names, being and one, must be, so that being is more than one (*Soph.* 244b–e). The Stranger goes on to explain that the relationship between name (ὄνομα) and thing (πρᾶγμα), exploited by Odysseus, entails the multiplicity of being and concludes from this that to experience unity is necessarily to experience it as a whole of parts. Parmenides must make some concession to complexity, which in turn entails that being be treated dialectically; consequently, for the sighting of being to be sufficient and communicable to another, there is the greatest need of the dialectical science, of which the precise character is infamously obscure but its general character, as considering the complexity of being, is plain enough. What the dialectical science would accomplish, should we be so blessed to acquire it, is the apparently impossible synthesis of speeches tailored to the particular and so necessarily suboptimal position of one's audience, a sort of image that the Stranger refers to as an apparition (φάντασμα), and of speeches tailored to the beings, a sort of image that the Stranger refers to as a semblance (εἴκων). For, as the Stranger notes, the dialectical science allows one to discern what is and, in turn, to teach it to *others*. The Stranger thus returns, in the argument, to his decision to accommodate the others, in the action, by speaking in conversational exchange (with Theaetetus), rather than engage in an inhospitable monologue, which would necessarily be indifferent to the character of his audience. But in so doing the Stranger must present our grasp of being as necessarily mediated by images, which both are and are not the things imaged and so can only be understood by mixing being and nonbeing: that is, by killing father Parmenides.

Parmenides, however, or at least his goddess, does not seem to grasp the obstacle communication imposes on his, or her, teaching. In the context of his discussion of name and thing, the Stranger quotes a particularly revealing line from Parmenides's poem—to my mind, the most revealing in the extant fragments—a line in which the goddess describes being as "resembling the bulk of a well-rounded sphere" (εὐκύκλου σφαίρης ἐναλίγκιον ὄγκῳ, 8.43). What's so revealing in this context is the word "resembling" (ἐναλίγκιον), which appears to be an anticipation, voluntary or not, of the problem of images in the *Sophist*. The goddess breaks off a familiar part of being and uses it as an image for being as a whole, so as to make her revelation intelligible to the young man before her.[10] There is a tension, in other words, between her account of being and her didactic aim with the youth who has traveled to see her, the very tension exhibited in the Stranger's distinction between εἰκόνες and φαντάσματα.[11] Parmenides's goddess appears as blind as Polyphemus to the demands of being a good host.

Mind and Nature

There is, of course, a flaw in this interpretation of Socrates's allusion, in that Odysseus's encounter with Polyphemus doesn't end happily for Odysseus; the episode is rather the beginning of his woes. As he departs from Polyphemus's island, Odysseus—having doubled-down on his reliance on his μῆτις (see *Od.* 9.422)—cannot resist the temptation to declare his name, that of his father, and that of his home city openly, a revelation that in turn allows Polyphemus to beseech his father Poseidon to exact revenge on his behalf. Odysseus's desire to be known for his blameless mind ends up keeping him from home another ten years, long after the other heroes have returned home or gone to Hades. Odysseus comes to understand his mind's folly when, after enduring the loss of all his ships besides his own at the hands of the Laestrygonians, he comes to the island of the nymph Circe. Eurylochus leads some of the men to Circe, who transforms them into pigs. That is, all are transformed save Eurylochus, who returns in dismay to tell Odysseus what has happened. Odysseus immediately arms himself and departs for Circe's den. Along the way, Hermes stops him and reveals to him the antidote for Circe's charms, the moly plant. But Hermes doesn't just pluck the plant from the ground and hand it to Odysseus, he also "shows him its nature" (φύσιν αὐτοῦ ἔδειξε, *Od.* 10.303).[12] That is, Hermes imparts knowledge of nature to Odysseus. After Circe's attempt to turn Odysseus into a pig fails, she inquires, as Polyphemus before her, into who Odysseus is (*Od.* 10.325). She then remarks that "not one (οὐδέ τις) other man has endured these poisons" and so infers from this that Odysseus possesses "some unenchantable mind (νόος)" (*Od.* 10.327–29). Whereas in the episode with Polyphemus, Odysseus's blameless mind gave him sophistic abilities, here Hermes's instruction makes his mind beyond enchantment, that is, beyond such enchantments as a blameless mind might devise. Hermes thus completes the education of Odysseus that had begun in Polyphemus's cave, bringing him from awareness of the manipulability of appearances to knowledge of nature.

Parmenides echoes this relationship between mind, nature, and appearance in the fragments detailing his cosmology. At the center of this cosmos is a goddess (δαίμων) who helms all things (12). But for there to be anything at all, the δαίμων must "first, among the gods, devise eros (ἔρωτα . . . μετίσατο)" (13). When nature does eventually come up, it is in relation both to natural growth and a sort of necessity governing motion (10). Parmenides thus presents the growth and necessity of things as having an order intelligible to mind, μῆτις. Correlatively, the Eleatic Stranger hypothesizes in the course of the *Sophist* the need for a dialectical science that renders its possessor capable of sorting the kinds as the man skilled in the art of writing sorts letters, so that he perceives adequately the relationship between looks, particulars, and wholes (*Soph.* 253d5–e2). Like Odysseus before them, then, both Parmenides and the Eleatic Stranger attempt to establish a relationship between mind and world that sorts out nature into natural kinds. In this sense, then, they stand closer to the Odysseus who overcomes Circe's charms than the Odysseus who declared his name proudly and stupidly before Polyphemus.

It should be remembered, however, that Parmenides's goddess in fact condemns this path, characterizing the attempt to sort out the natural order or cosmos as the way of know-nothing mortals, who wander around two-headed (6.3–5). This way involves the very mixing of being and nonbeing that she forbids us, bringing the apparent peak of Parmenides's poem closer to the Cyclops, in its blindness to the relationship between being and nonbeing (8.51–59). Plato's Homeric critique of Parmenides would thus suggest that the actual peak of Parmenides's poem is rather its cosmology than its ontology, in that the former prefigures the Stranger's willingness to relax the goddess's argument and allow for some mixture of being and non-being. The Stranger resembles Odysseus, in that first he must escape the Cyclops Parmenides, so that later he can obtain such knowledge of nature, through the dialectical science, as would make him and Theaetetus impervious to the sophist's charms.

Hades

Yet this cannot be the whole story. For, again, Odysseus's travels don't end here, at Circe's island, but take many years and cover a great distance. This is likewise true of the Stranger, whose work is not done with the sophist just because the *Sophist* has ended. Despite all the guises of the sophist that he and Theaetetus have encountered, they have neglected the primary or at least most conspicuous of his guises, that of the statesman or politically skilled man (πολιτικός). When the Stranger and Theaetetus listed the sophist's claimed areas of expertise, the Stranger asked, "But what, in turn, about laws and all the political things, do [sophists] not promise to make [their students] disputatious [about these]?" Theaetetus pregnantly responded, "Well, no one would converse with them, in a word, if they didn't promise this!" (*Soph.* 232d1–4).[13] To investigate the sophist adequately, then, one must proceed through his foremost apparition as a statesman.[14] Nowhere in the *Sophist*, however, does this guise come up. Rather, it is not until the *Statesman* that we get anything like a thorough treatment, when, in its latter portion, the Stranger sets as his and Young Socrates's task separating the scientific statesman from all the partners in other regimes, those he calls the greatest sorcerers and charlatans of all the sophists and the most experienced in this art (*Stat.* 267c5–268d4, 291a1–c8, 303b8–c7, esp. 291c3–4 with 303b8–c5). Is there some sense in which the continued travels of the Stranger parallel those of Odysseus?

Odysseus spends a year on Circe's island, enjoying the fruits of his "unenchantable mind," possessed of knowledge of nature; a year now spent, his men interrupt his repose to remind him of his fatherland (πατρίδος αἴης, *Od.* 10.472). It seems that Odysseus's transformation into knower of nature has disconnected him from the all-too-human attachment to the land of one's origin. Is Parmenides's goddess likewise too quick to transcend and, from on high, dismiss man's all-too-human roots? Encouraged by the reminder of his fatherland and, in addition, of the prophecy of his eventual return (*Od.* 10.473–74), Odysseus tells Circe he would like to return home (οἴκαδε, *Od.* 10.483–86); in response, she informs Odysseus that he must travel to Hades to speak with the blind prophet Teiresias, who alone among the shadows has intelligence or a mind (νόον, *Od.* 10.488–95). When, after his terrifying travels, Odysseus finally encounters Teiresias, the seer informs him of the conditions in Ithaca and permits him to punish his wife's suitors as he sees fit; but he further instructs him—and this lesson is quite central and easily passed over—that afterward he is to appease Poseidon by taking his oar and traveling so far inland that he meets a people ignorant of the sea, the sign of which being that they will mistake his oar for a winnowing fan (*Od.* 11.100–34). Odysseus is to establish justice in his house and then, in an apparently unrelated task, to honor Poseidon by adding to his worshippers those who would otherwise have no reason to know of him, let alone revere him. For this very reason, however, these worshippers will understand Poseidon not as the god of the sea but as a god who separates the wheat from the chaff. That is, they will understand him as a punitive God concerned not with avenging the harm Odysseus did his son Polyphemus, but rather with punishing and rewarding the deeds of men in the afterlife. In other words, Odysseus is to learn that what holds the political community together is not a blameless mind or a mind beyond enchantment, no less than the private vengeance of Poseidon or the personal pleasure of his son Polyphemus; rather, it is justice and the prospect of divine retribution.

A similar obstacle confronts the Eleatic Stranger in his search for the statesman. The Stranger, accompanied by his interlocutor Young Socrates, presents statecraft as a science, specifically a cognitive science that gives orders or injunctions to others with an eye to the good of human beings. Unlike other artisans, however, the statesman finds his rule challenged not just by "the tribe of heralds," among them the priests (*Stat.* 289d3-4, 291a1-2), but by "the greatest enchanter of all the sophists and the one most experienced in this art" (*Stat.* 291c3-4). What separates the statesman from this enchanter is, it turns out, his political knowledge (πολιτική ἐπιστήμη), which necessarily belongs to a few and proceeds lawlessly, that is, in potential violation of the laws; "a law," the Stranger observes, "could never enjoin the best, while at the same time comprehending precisely both what is best and what is most just for all," on account of "the dissimilarities among human beings and actions" and because "never is anything of the human things, in a word, at peace" (*Stat.* 293e7-294b4). At the same time, however, the statesman, who is in a certain manner (τρόπον . . . τινὰ) a lawgiver (*Stat.* 294a6-7), could never be sufficient (ἱκανόν) to give injunctions to all the crowds while also giving what is suitable to each precisely (*Stat.* 294e9-295a3). In other words, the statesman with precise knowledge must rule using a necessarily imprecise means, law, which is a mere imitation of his knowledge. With the entrance of imitations comes the sophist, who takes the form of the various nonscientific regimes but most threateningly the tyrannical imitator of the scientific statesman, both of whom would, on occasion, prefer to do something better than what is prescribed by the laws, albeit for different reasons (*Stat.* 301b10-c5). The solution to this problem is the rule of law or the inviolability of the ancestral things, the things of one's fathers (τὰ πάτρια), on pains of death and "all the extreme things" (*Stat.* 297d4-e5). What the Stranger means by "all the extreme things," apart from death, is unclear. The only time τὰ πάτρια have been mentioned prior to this point was in the discussion of the religious imitators of the scientific statesman (*Stat.* 290e7), which suggests that "all the extreme things," apart from death, might be religious in character, that is, the sufferings of the soul in the afterlife. If this is indeed the Stranger's suggestion, then like Odysseus he advances the claim that what preserves the community is not the wholly cognitive science of statecraft but law and the prospect of divine retribution. Just as Odysseus listened to his men to return to his fatherland, so too the Stranger eventually acknowledges the scientific merits of perhaps the most antiscientific of allegiances: the allegiance to the ancestral things or τὰ πάτρια.[15]

Antinous and Penelope

At this point, it is helpful to introduce the second passage to which Socrates alludes at the outset of the *Sophist*.[16] Upon returning to his fatherland, Odysseus finds himself in a similar situation and with a similar intention in mind to when he visited Polyphemus. Again hiding who he is but this time in the guise of a beggar, Odysseus goes into his palace, stirred by Athena,

> that he might gather bread from each of the suitors
> and he might know which ones are righteous and which
>   lawless. (17.362–63)

Most of the suitors oblige this beggar, save for Antinous, who reproaches him severely and eventually strikes him with a stool. It is in response to this savage act that the other suitors reproach Antinous:

> Antinous, not nobly did you throw at the miserable vagrant,
> accursed you, if indeed he is some god on high,
> and the gods—assuming the likeness of strangers of another
>   land,
> coming to be of all sorts—frequent the cities,
> observing the arrogance and good governance of human
>   beings. (17.483–87)

Reflecting on this and other abuses later in the night, Odysseus will eventually recall what he endured in Polyphemus's cave, tempering his desire for vengeance until the next morning (20.1–30). Odysseus's encounters with the Cyclops and the suitors thus bookend, within his larger travels, his journey from hubristic yet clever deceiver, to knower of nature, and finally to avenger and enforcer of right.

Odysseus's restoration of justice to Ithaca involves the punishment not just of Antinous, nor of the other suitors, but of the serving women, as well.[17] While their execution appears peripheral to Odysseus's great act of vengeance against the suitors in Book 21, it is upon hearing them joyfully carouse with the suitors that Odysseus is reminded of what he endured at the hands of Polyphemus. Why does their transgression in particular anger Odysseus? Between Antinous's assault of Odysseus and Odysseus's recollection of the Cyclops, Odysseus encounters and is abused by the serving woman Melantho, who scoffs at the notion of attending to Penelope, instead sleeping with the suitor Eurymachus. And she does this despite the motherly role Penelope has played in her life. Later that evening, Odysseus again encounters her and receives abuse, this time in the presence of Penelope, who chastises her for her treatment of their guest. When, thereafter, Odysseus (again, in disguise) and Penelope introduce themselves to one another, Penelope relates the memorable tale of how she has maintained order in her husband's absence. Under divine instruction, she told the suitors she will not take any of them in marriage until she finishes weaving a funeral shroud for Odysseus's father Laertes. All agree. By day she would weave, fulfilling her agreement, while by night she would undo it, in violation of the same. Her ruse fell apart, however, when her serving women eventually exposed her. To preserve order, then, Penelope was forced to violate her agreement; her willingness to do so, despite the attendant risks, proves her political wisdom and commitment to the good of Ithaca and her family. Nevertheless, her transgression, necessary as it may have been, invites that of the serving women, as well.

Penelope thus experiences the very problem about which the Stranger warned, that the scientific statesman must occasionally depart from law to accomplish what is best, yet in so doing invites the sophistry of his tyrannical imitators. The serving women's transgression was thus of the highest order, in that their lawlessness initiated the complete degeneration of political order in Ithaca. Justice cannot be restored to Ithaca without their being punished as harshly as (and less ceremoniously than) the suitors. Picking up on this second allusion, as he had the first, the Stranger presents the scientific statesman as a weaver, who plaits together the souls of the spirited and the gentle by issuing inviolable laws that soften the one and harden the other. Socrates's twin allusions to the *Odyssey* thus guide the Stranger, a lapsing Parmenidean, in what we can call Plato's Homeric critique of Parmenides. The substance of this critique is that the goddess's prohibition against speaking nonbeing makes any understanding of the political realm impossible, where the sophist and his spoken images run wild. And though inviolable law offers a way to put the problem of sophistry to rest, the Stranger acknowledges that law is but a second sailing, a second-best solution, and thus a reluctantly accepted imitation of the scientific statesman's prudent rule. But perhaps worst of all, as we have noted, the goddess herself employs the language of imagery and thus unwittingly betrays how much her own reasoning is a creature of the human realm, just as sophistry and law. To return to the original issue, this means that the philosopher necessarily appears as sophist and statesman and, consequently, that the Stranger is much more of a Socratic than he lets on. Gracious to his host, the Stranger never reveals to Theodorus his affinity for Socrates but like Odysseus hides what he is.

Parmenides and Plato

In developing Plato's Homeric critique of Parmenides, subtly proposed by Socrates and elaborated by the Stranger, we have done Parmenides the injustice of denying him any defense. And there is much room for defense, since we have noted that Parmenides distances himself from the arguments attributed to him by stating them not in his own name but rather putting them in the mouth of an anonymous goddess.[18] Parmenides further distances himself from the goddess by portraying her disposition toward her audience as shifting from argument to argument. At first, her tone is merely didactic, indicating her intention to instruct the youth before her and carry off her instruction to others (2.1). Elsewhere, she is firmer, commanding her audience to perceive that at which her arguments have aimed (4.1). A bit later, perhaps out of frustration at her audience, her tone is more reproachful and contemptuous, as she describes mortals as knowing nothing, wandering two-headed, their perplexity leading their minds to wander off, so that they are effectively blind and deaf, a tribe without judgment (6.4–7).[19] Still elsewhere, she laments how her argument is contested, blaming habit and experience for compelling human beings to have unseeing eyes and resounding ears and tongues (7.3–6). All such signs of struggle disappear, however, during the cosmological fragments. Parmenides, then, appears to highlight the problem of communicating his ontology.

In Plato's defense, however, he seems well aware of the difference between Parmenides and the goddess. In the *Parmenides*, the aged Eleatic submits his hypothesis, that unity is, to thorough critique in an eightfold set of deductions from both the positive and negative versions of the hypothesis, that is, if unity is and if unity is not. In the first of these deductions, Parmenides argues that, if unity is, it can in no way be many. But if it is in no way many, then it cannot be a whole with parts. That ultimately leads to the conclusion that such unity cannot be known nor even spoken of. He thus concludes that this cannot be the way things are and goes on to consider, in the second deduction, such unity as is mixed with plurality, a whole of parts that is ultimately intelligible to human beings. The thrust of the argument suffices to show, at least preliminarily, that the deductions are probing rather than fixed and routine. That is, Parmenides seeks to test various possibilities, moving on from one to the next when the path taken proves untenable or an alternative proves in need of exploration. In short, Parmenides is not the dogmatic monist he is often taken to be. We tend to identify Parmenides with his goddess, as did many in Parmenides's own time, though Plato seems to have distinguished them and so distinguished Parmenides's own thought from the much more dogmatic school that arose around him.

The relationship between thinker and school informs the drama of Plato's *Sophist*, as well. Theodorus informs Socrates that the Stranger is "a comrade of those around Parmenides and Zeno" (*Soph*. 216a3). As we later learn, the Stranger has long had his problems with Parmenides's teaching; as we saw, however, from Plato's *Parmenides*, Parmenides himself took issue with the teaching in his poem. We are inclined, therefore, to distinguish the Eleatics into three groups: Parmenides and Zeno, the center around which the others turn; the dogmatic devotees to the teaching of Parmenides's goddess; and those like the young Socrates and the Stranger, who find Parmenides's teaching problematic and so depart from it. By deviating from the dogmatic Eleatics, the Stranger in fact comes closer to Parmenides's own thought, the same thought he once imparted to the young Socrates.[20] The irony is that, in breaking with the dogmatic interpretation of Parmenides and the Theodoran interpretation of Socrates, the Stranger ends up coming closer to both and, thereby, closer likewise to Homer.[21]

Olympus

We can now look more closely at Parmenides's own allusions to Homer.[22] Features of the poem, especially the journeying motif, draw the youth close to Odysseus and specifically to his eventual journey to the underworld to meet Teiresias.[23] The image of the chariot, however, recalls Hera's attempted journey to her husband Zeus atop Olympus. The competing images for the

Olympus as Hades | 73

final destination of this journey—Olympus and Hades—could not be any more opposed. As Alexander Mourelatos has observed, "The topography of the journey is blurred beyond recognition."[24] Since we have dwelled at such length on Odysseus's travels, let us look briefly at the journey in the *Iliad* to which Parmenides alludes: Hera's trip to Zeus atop Olympus in Book 5.

Amid Diomedes's impressive display, aided by Athena, the Argive Tlepolemus goes up against Sarpedon, the son of Zeus. Sarpedon kills Tlepolemus, but not before the latter throws his spear and wounds Sarpedon in the thigh. Saved by his father Zeus, Sarpedon lives to fight another day. Hector, encouraged by Sarpedon and aided by Ares, answers Diomedes's work with a slew of killings of his own. This worries Hera so much that she mounts a chariot with Athena, clad and ready for war. Hera arrives and requests permission from her husband to wound Ares. Zeus obliges; but on the following morning he commands the other gods not to help either the Trojans or the Argives, backing up this command with a statement of his near universal power. Not much later, Hera and Athena test Zeus's threat and again mount a chariot against him. Homer invites us to compare these passages, repeating lines in both while varying the circumstances. While the preparation of the chariot and the arming of Athena are familiar, this time Zeus sends Iris to warn them of the consequences they will endure, should they follow through with their contravention of his order. The goddesses prudently leave off from their attack. When, in turn, the gods gather on Olympus, Zeus reveals his plan to allow Hector to reach the Achaeans's ships and slay Achilles's companion Patroclus. By the end of Book 12, everything is in place: Hector has reached the ships, and Patroclus is eager to return to battle. Zeus thus withdraws at the beginning of Book 13, confident that everything will unfold as planned.

Zeus's confidence proves poorly placed, for Hera makes a third trip against him. This time, however, she comes prepared not for war but for lovemaking. Her success puts Zeus in a postcoital slumber that allows her and Poseidon to aid the Argives more openly, until Hector is wounded and drawn down from his ambitions. To accomplish his plan, Zeus must encourage Hector anew, and for this the killing of Patroclus is necessary. But because Hector has pulled back to Troy, Zeus must first encourage the reluctant Patroclus to violate Achilles's advice to stay near the ships and instead attempt to take the enemy city. For that, an encouraging kill is required. Here Sarpedon enters the story again, only now Zeus cannot save him but must, lest Hera and the other gods openly rebel once more, let him die. Zeus, who had attempted to make all one with his will, is forced by Hera's erotic (as opposed to martial) attack to make concessions to the lesser gods and men. We sense, then, in Parmenides's allusion the subtle suggestion that the Olympus atop which the goddess is supposed to reside is not so unified and impenetrable a place as she would have us believe.

Olympus as Hades

Parmenides thus brings his two opposed images, of Olympus and Hades, closer together and so invites us to consider his teaching in light of the similarities between Odysseus's travels and Zeus's suffering. The common element they share is a love of their own, be it fatherland, family, or son. Zeus's sacrifice, necessary for him to maintain unity, ultimately humanizes him, which wouldn't be such a bad thing if he weren't a god. Homer's powerful image for a humanization that betrays his divinity is to have the king of gods and men cry tears of blood at the death of his son, Sarpedon. It is a strange thing for a god to do, since he could always just have another child, or many children over, as indeed he has (compare *Iliad* 3.441–46 with 14.314–27). Zeus learns too late what Odysseus learns just in time: the connection between mind and the world of particulars, of particular cities, homes, wives, and children. That is, Zeus would have done well to venture into Hades himself and learn from Teiresias, as Odysseus eventually will, that mind rules men not purely but through justice and just punishment.[25] The true peak of the Homeric epics is not Olympus, where Zeus dwells, but Hades, where Odysseus ventures.

It is worth mentioning a final connection between Homer and Parmenides, one that has not received any attention, at least to my knowledge. At the center of Zeus's fall in the *Iliad* is the experience of love (ἔρως), the passion that leads to the birth of his son Sarpedon and that allows Hera to undermine his initial plan. Zeus even goes so far as to mention the erotic longing that led to the birth of his greatest children—Sarpedon is notably absent from this list—as he's in the grips of erotic longing for Hera (see, again, *Il.* 14.314–27). That Zeus has such an experience of love is the chief and, I think, decisive bit of evidence in the case against his divinity, since love speaks to a troublesome finitude that is simply all too human. Homer thus suggests that the gods are best understood in terms of the human soul and its longings, a shift he accomplishes in moving from the *Iliad* to the *Odyssey*, a shift likewise echoed in Parmenides's turn from being to cosmos.

How so? Earlier, I proposed that the true peak of Parmenides's poem is to be found in his cosmological fragments, as opposed to the apparent peak in the ontological fragments. The cosmological fragments accommodate man, while in the ontological fragments the goddess struggles to win her audience to her account of being as one, removed as it is from human speech and experience. In the cosmological fragments, the revelatory goddess (θεά), no longer exhibiting signs of struggle, speaks of a governing goddess (δαίμων), whose first task in bringing order to things is devising love (ἔρωτα . . . μετίσατο, 13).[26] Parmenides elucidates the human experience of this order in two further fragments. The first claims that pregnancy on the right side (δεξιτεροῖσιν) produces boys and on the left side girls (17). A rational distinction within experience promises the satisfaction of what's preferable (τὸ δεξιόν, cf. 1.3: δεξιτερήν). But elsewhere Parmenides discusses vexatious birth defects, in what amounts to a return to the unpredictability of the particular (18). At the heart of Parmenides's poem, then, is an exhibition of the tensions between spirited anger (θυμός), erotic longing (ἔρως), and mind (μῆτις). Platonic as this interpretation of Parmenides might seem, we are saved from accusations of anachronism by Plato's admissions of Homer's influence on his analysis of the human soul in the *Republic*, to say nothing of the patently Parmenidean imagery of his account of the soul in the *Phaedrus*. Attempting to reconcile these two images, in light of their poetic and philosophic forebears, would take us deeper into the question of the subtle, even subterranean interplay between Homer, Parmenides, and Plato. It is a question that I hope the preceding convinces others to pursue.[27]

Notes

1. See the parallels catalogued by Mourelatos 2008, 4–11; as well as Coxon 2009, 7–18.

2. Brann 2001, 123.

3. See the Parmenides and Plato section, first paragraph.

4. Socrates plays on Theodorus's phrase κατὰ τὴν χθὲς ὁμολογίαν with κατὰ τὸν Ὁμήρου λόγον (Plato, *Soph.* 216a1, 6). There follows shortly thereafter a second bit of wordplay, again by Socrates, on Ἐλέα and ἐλεγκτικός (*Soph.* 216a3, b6). For my interpretation of Homer's *Odyssey*, I am deeply indebted to Benardete (1997).

5. Ronna Burger first kindled my interest in these allusions; in her graduate courses on Plato and Aristotle, she would often remark how much their allusions to the poets were the best instruction on how to read them, particularly Homer. The present chapter is an attempt to work out a parallel between the core philosophic issues of the *Sophist* and *Statesman*—being and nonbeing, on the one hand, and law and weaving, on the other—and the passages to which Socrates alludes—Odysseus with the Cyclops and Odysseus with Antinous and Penelope. I'm grateful to Ronna for her helpful comments on a draft of this paper.

6. All citations of ancient Greek texts are to the Oxford Classical Texts editions, except for the citations of Parmenides, which are to Coxon 2009, but according to the traditional Diels-Kranz numbering. All translations are my own.

7. On the location of Polyphemus, versus the other Cyclopes, see *Od.* 9.188–89, 399–400.

8. Note that Socratic dialogue is not a broken-up monologue, as is the Stranger's, but the examination of perplexities hiding among our opinions.

9. It's reasonable, I think, to translate μῆτις as "trick" here. One could say that, in the Cyclops episode, Odysseus's mind operates on the level of trickery as opposed to how it operates on Circe's island or back in Ithaca.

10. To what among the beings does Parmenides mean to compare being? Consider what Parmenides says of the moon in Fragment 11.

11. It is even tempting to see in the Stranger's quoting of this line a playful reference to the Cyclops's single, circular eye. For it fits nicely with the image of the Stranger as an Odysseus who escapes from the trap, in which Parmenides's rounded being has placed him, by grasping the relationship between Odysseus's blameless mind and the deceptive names that mind employs. See the previous note.

12. Many have noted that this is the first use of this key philosophic term in extant Greek literature (and the only occurrence in Homer). What I hope to develop here is the connection between this occurrence of φύσις and the sustained emphasis on μῆτις throughout the *Odyssey*.

13. Consider, too, 225c7–10 and 268b1–c4.

14. See, e.g., Protagoras's attempt to augment what the gods have given us in his great speech in Plato's *Protagoras*, which should be read in light of the central lesson of the next paragraph.

15. For this paragraph, I am indebted to an excellent paper delivered by Seth Appelbaum at the 2015 meeting of the Northeastern Political Science Association, with the title "Law and the Restriction of Statesmanship and Philosophy in Plato's *Statesman*."

16. One important episode in Odysseus's travels that I simply do not have space to discuss is what he learned during his stay on Calypso's island. Bolotin notes that "the Odysseus who returned home from Calypso's island seems to have learned the secret of a good human life, a life of contentment with his fate and

of whole-hearted communion with his wife and family," though he goes on to complicate this picture (Bolotin 1989, 55).

17. When Odysseus, still in disguise, kills Antinous, the rest of the suitors scold him angrily. Odysseus responds by revealing himself and leveling three charges against the suitors:

> Dogs, never did you think I would come home (οἴκαδε), returning,
> from the land of Troy, since you wasted my home,
> forcibly laid beside the serving women,
> and slyly sought to marry the wife of me, though still living. (21.35-38)

The second of the three charges is puzzling, since eventually all the serving women will be put to death for having slept with the suitors; likewise, the revelry of the night before appears entirely consensual. Certainly Odysseus perceives it as such. Odysseus thus appears to trump up the charges against the suitors by temporarily absolving the serving women of their guilt. But to what end? Eurymachus responds to Odysseus by trying to place blame on Antinous for seeking to marry his wife, adding to the charges a plot to kill Telemachus. Odysseus, however, reiterates his intention to punish *every* suitor for his transgression, a verdict possible only on the basis of the rape charge. What's most necessary is to purge Ithaca of general lawlessness, even if the particular actions do not merit death. See the remainder of the paragraph, above. Compare Bolotin's remarks on Leiodes in Bolotin 1989, 51-52.

18. Points in this and the following sections are supported in greater detail in Priou 2018.

19. It's a wonder that she speaks to mortals at all.

20. A more difficult question is whether the Stranger himself was aware of this phenomenon. Socrates's remark that he witnessed Parmenides speaking in a dialogue rather than a monologue might suggest to the Stranger that Parmenides had, on some level, grappled with what the Stranger comes to formulate as the tension between φανταστική and εἰκαστική.

21. Regarding the last two points, see the Mind and Image section, paragraph 2.

22. In Priou 2018, I attempted to spell out the drama of Parmenides's poem, or what we have of it. At the time, however, I did not touch on how Parmenides's allusions to Homer inform this drama, and as a result certain fundamental features of his thought escaped my notice. I hope to have corrected for that deficiency here.

23. See Havelock 1958, 136-43.

24. Mourelatos 2008, 15; the problem is distilled at 12-16. One particular reflection is helpful here: "We are tempted to say that this [blurring] is an effect of the 'abbreviated-reference' style, and that if we had more of the early epic, we

would be able to decipher the references. This is possible. But a hypothesis closer to the facts at our disposal is that the blur is intentional. After all, Parmenides could write with shattering clarity and precision when he wanted."

25. On the questionable justice of Zeus, as opposed to men's expectations of him, see Ahrensdorf 2014, ch. 1, "The Theology of Homer."

26. For a fuller discussion of the cosmological fragments, see Priou 2018, 191–95.

27. A still further question, beyond my understanding, is the relationship between Parmenides and Hesiod, whose influence on Heraclitus is likewise not often enough appreciated; all three converge on the relationship between night and day. See Miller 2006, 7–9, 18–28. They are principally close on the high (though by no means equivalent) place each gives love (ἔρως). Plato also recognized this (see *Symposium* 178a8–c3).

Section II
Truth

Chapter 4

Parmenides's Fragment 2 and the Meaning of Einai

Colin C. Smith

Parmenides's poem is the story of one mortal's quest to make contact with the deepest and most elusive of divine revelations that is also, paradoxically, the most ubiquitous of notions: being itself. Similarly, my chapter is a story about the trip we collectively have taken in trying to grasp the meaning of the narrating goddess's provocation in what we now call Fragment 2. The question concerns the goddess's use of "is" (ἔστι and ἔστιν) and "to be" (εἶναι, 2.3 and 2.5), henceforth collectively "the '. . . is . . . ,'"[1] to describe a sense of being that must be and cannot not be (2.3), which entails a path of inquiry that contrasts with the "is not" (οὐκ ἔστι and μὴ εἶναι) that can serve experientially as a guide for our understanding. Much has been said about it by commentators, but no account of the ". . . is . . ." has yet exhausted this rich subject.[2]

My aim in this chapter is to return to this much-discussed issue in Parmenides scholarship to tell its story, reflect on our collective grasp of the goddess's insight, and offer some further points that might contribute to it as well. My own view of the meaning of the ". . . is . . ." that I develop is neither entirely new nor, so far as I can tell, entirely redundant with any other interpretation that has been offered. In short, I take my cues from the goddess's connection of being to knowing (νοῆσαι, 2.2, and elsewhere) and prohibition against the possibility of what-is-not "*as such*" (γε, 2.7) to develop a kind of "two-term" reading that concerns the necessity of determinately intelligible predicative structure entailed in all beings through what I call "being-as-such." On such a reading, the goddess teaches us to understand being through seeing how inquiring into the nature of beings allows the beings to show themselves to thinking, while guiding us to the ubiquity of being by indicating the impossibility of nonbeing as such.

My account will be intertwined with a discussion of this fragment's historical reception in the scholarship, drawing upon the major lines of argumentation while also offering some friendly emendations. According to my account, the goddess's target is not, as others have argued, so-called existence, not exactly the definitional nature of thinghood, nor the modal sense of necessity: though all these issues are closely related. Instead, the notion identified in Fragment 2 is the principle of ontological grounding, the most basic at least in a sense, that is itself simple and unique, and necessarily implies complexity as a binding relation. On my reading, the task of Fragment 2 is to call this simple and ubiquitous notion of bindingness to thought before developing its further specificity, including its identity with thinking as mentioned in Fragment 3 (discussed by Michael Wiitala in his chapter herein) and demand for further specificity in terms of τό ἐόν, or "that which has being," developed in Fragment 8 (and discussed by Eric Sanday in his chapter). The goddess here opens the space for a new kind of understanding of reality through the paradigmatic notions of abstraction, elision, commonness, and unity, further framed by the notion of necessity, the function of knowing, and the impossibility of "is not." I submit that the poem depicts the mortal struggle to hold on to this insight into being as it comes into and then slips out of focus, and the ways that we might habituate ourselves to return back to it and preserve it as best as we mortals can.

We proceed as follows. Before considering Fragment 2 itself, we take a brief detour in the next section to consider the understanding of being in the history of philosophy and some reasons why our thinking has become muddled to preclude our firm grasp of the Parmenidean insight into being. After that, we turn to Fragment 2, and I offer a translation and highlight a few textual issues that will be central to the interpretation that follows. I then embark upon summaries of the major lines of interpretation of the ". . . is . . ." speaking to what I take to be the strengths and imperfections of each. I conclude by developing a cumulative account of the Parmenidean meaning of being, which I call "being-as-such."

The History of Understanding Being

The philosophical orthodox view of being, such as it was, developed significantly in the ancient Greek lineage after Parmenides through some points that Aristotle makes in his *Categories* and the ways they were subsequently taken up. Here, Aristotle distinguishes between being in two senses: that which is "of itself" and that which is "of something further" (the latter of which Aristotle further subdivides into nine categories that we need not consider here). For example, consider a set of things that *are*: Socrates, Socrates's snubnosedness, Milo the cat, Milo's affectionate nature, my paperweight rock, and the rock's locatedness on my desk. Within this set, Socrates, Milo, and the rock all *are* only with respect to themselves and nothing further. Although such terminology does not map onto the ancient Greek vocabulary, we might say that Socrates, Milo, and the rock all "exist independently." By contrast, examples like Socrates's being snubnosed, Milo's being affectionate, or the rock's being located *here* on the desk are dependent properties, or instances of being that could not *be* without more basic entities like Socrates, Milo, and the rock.

Tabling the question of whether Aristotle actually held this view himself,[3] from such a distinction follows the view that being is most essentially being an independent entity like Socrates, Milo, or the rock. Such was later called "substance ontology," and regardless of whether this in fact captures Aristotle's own understanding, the influence of substance ontology upon subsequent medieval and early modern philosophical thinking was enormous:[4] for many generations of philosophy, "to be" was understood most primarily as to "exist independently."

Significant in the story of our collective journey is the new chapter that began in the nineteenth century with the identification of an apparent problem implied by such a view of substance ontology, understood very broadly and imprecisely. At this historical juncture, Anglo-American philosophers like John Stuart Mill began insisting upon a fundamental divide between senses of the verb "to be." For example, Mill writes:

> Many volumes might be filled with the frivolous speculations concerning the nature of Being, (τὸ ὄν, οὐσία, Ens, Entitas, Essentia, and the like) which have arisen from overlooking [the] double meaning of the word 'to be'; from supposing that when it signifies to *exist*, and when it signifies to *be* some specified thing, as to *be* a man, to *be* Socrates, to *be* seen or spoken of,

to *be* a phantom, even to *be* a nonentity, it must still, at bottom, answer to the same idea [. . .]. The fog which rose from this narrow spot diffused itself at an early period over the whole surface of metaphysics. Yet it becomes us not to triumph over the great intellects of Plato and Aristotle because we are now able to preserve ourselves from many errors into which they, perhaps inevitably, fell. (1843, I.iv.1)

Mill here indicates what he takes to be a fundamental divide among the senses of being and the errors yielded by a failure to recognize this divide. Most primarily, this division supposedly entails marking off the "existential" sense as a "one-term" use of the verb "is" (e.g., "Socrates is" in the sense meaning "Socrates exists") from other, "two-term" uses. Examples of "two-term" senses include the predicative (where "is" acts as copula and picks out a feature of the subject, e.g., "Socrates is snub-nosed"), veridical (where "is" emphasizes truth value, e.g., "'That Socrates is a philosopher' is true"), identity-marking (where "is" picks out the most essential and defining features that mark off the subject from all others, e.g., "Socrates is the son of Phaenarete and Sophroniscus"), the existential-locative "is" (where the "is" picks out some combination of existence and location implied by the *there is*, e.g., "*There is* this guy in Athens who is a real gadfly"), and others besides. Mill's is not exactly the distinction to which the substance ontologist is committed following the interpretation of Aristotle's *Categories*, but the affinities are clear. In both instances, being is understood as most basically entailing what we would call "existence," and all other senses of being are taken to be secondary and derivative.

These alleged distinctions and problems raise the question concerning the meaning of historical articulations of the verb "is." The ". . . is . . ." of Parmenides's poem, to which we will turn momentarily, thus became a prime target of debate.[5] In short, the assumption was that Parmenides must have meant one (or more) of these disparate senses, and the task fell upon the commentator to argue for which modern sense(s) cohere(s) most compellingly with the way Parmenides has the goddess describe being in the poem. Because Parmenides's goddess describes being as "whole" (8.4), "one" (ἕν, 8.6), and "undivided" (οὐδὲ διαιρετόν, 8.22), we have reason to think that Parmenides's object is not the divided notion of the substance ontology and Mill. At best, we might take the ". . . is . . ." of Fragment 2 to pick out one (or more) of their divided senses, marking a discovery that would later be "clarified" by further philosophical work; at worst, it might turn out that Parmenides "fell" into one or more of the many "errors" by failing to understand such distinctions and "overlooking [the] double meaning" as Mill describes.

These problems are partially due to the fact that the English verb "to exist" derives from Latin and has no clear correlate in Greek.[6] Whereas English speakers have this separate "exist" verb to use in cases indicating what the substance ontologist and Mill both take to be the "primary" sense of being, the ancient Greek speaker would need to use "to be" (εἶναι and cognates) in all cases, including what counts as primary and secondary for both the substance ontologist and those following Mill. Under the assumption that Parmenides must have had in mind one (or more) of these senses or otherwise have failed to distinguish them, frequently with concern over the lack of the "exists"/ "is" distinction in ancient Greek, there arose the wide swath of competing interpretations that we will consider below.

With this story in mind, I propose that we move further backward in the history of thinking of being to a time before substance ontology and divided senses of being were taken for granted. This will be valuable in no small part because substance ontology and the view of divided being face numerous problems. For instance, substance ontology gives credence to the false myth of independent existence. Nothing is truly independent: Milo the cat, for instance, depended on his cat parents for birth, and continues to depend on food, air, the justice of his cat roommates, and so forth; without this web of interconnected entities, Milo could not in fact "exist"—which here simply means "live"—and his seeming "independence" is in fact shown to be more complicated than it seems at first blush upon further scrutiny. We could tell similar stories for the other allegedly "independent" entities: the rock, for example, depends on the living things and nonliving forces in its environment not to smash it and erase its determinate unity by turning it into multiple rocks (and, if we are willing to posit such entities, it might depend further on the natures of "roundness" and "rockness" to be as it is). In any case, note furthermore that the substance ontologist's emphasis on "independence" is closely connected to the notion of creatio ex nihilo, implying as it does a being that does not ultimately derive from anything else.

Furthermore, if we take being to imply existence most primarily, we face numerous philosophical conundrums. For example, it seems that future generations do not "exist," and because (on such a view) nonexistent entities do not "have" anything due to lacking being in the allegedly basic sense, it follows that future generations do not have the right to a clean and healthy planet. This is ethically abhorrent and seems metaphysically problematic. Similarly, fictional entities like Superman do not "exist," and yet I can apparently make true assertions about them, like, "Superman is Clark Kent." This seems logically problematic. We should also note that this sense of "existence" seems to entail slippage between two separate notions: (1) being something instead of absolutely nothing (which, e.g., Superman, as a fictional entity, is) and (2) being an embodied thing in space and time (which, e.g., Superman is not), thereby needing disambiguation between these two senses.

I cannot here *prove* that these conundrums are ultimately decisive philosophically; but they are significant enough to illustrate a few reasons why we might want to revisit substance ontology and the view of divided being posited by Mill. While Mill implies that we are in a heightened position from which we might "correct" the ancients, I begin us upon the path that follows from the hypothesis that the Parmenidean view of being might help us to learn to overcome our own erroneous mortal thinking and its historical baggage, thereby having a better grasp of the meaning of being. Thus, on the presently hypothetical assumption that it is we who are confused and not Parmenides or the goddess, I turn our attention to the goddess's articulation of being to consider ways in which it might contrast with such views.

Fragment 2 Translation and Key Points

Here is the full text of Fragment 2 (following Gallop 1984):

> Εἰ δ' ἄγ' ἐγὼν ἐρέω, κόμισαι δὲ σὺ μῦθον ἀκούσας,
> αἵπερ ὁδοὶ μοῦναι διζήσιός εἰσι νοῆσαι·
> ἡ μὲν ὅπως ἔστιν τε καὶ ὡς οὐκ ἔστι μὴ εἶναι,
> πειθοῦς ἐστι κέλευθος (ἀληθείηι[7] γὰρ ὀπηδεῖ),
> ἡ δ' ὡς οὐκ ἔστιν τε καὶ ὡς χρεών ἐστι μὴ εἶναι, [5]
> τὴν δή τοι φράζω παναπευθέα[8] ἔμμεν ἀταρπόν·
> οὔτε γὰρ ἂν γνοίης τό γε μὴ ἐόν (οὐ γὰρ ἀνυστόν)
> οὔτε φράσαις.
>
> [Come now, and I will tell you, and you preserve the story
>   upon hearing it

> These are the only routes of inquiry there are for knowing
> The one, that [. . .] **is** [. . .] and that **is not possible** [for]
>   [. . .] **not to be** [. . .]
> Is the path of persuasion, for it attends upon truth,
> The other, that [. . .] **is not** [. . .] and that **is necessary**
>   (χρεών) [for] [. . .] **not to be** [. . .] [5]
> This I point out to you to be a path completely unlearnable
>   [alternatively, 'from which no learning comes']
> For you cannot know **what-is-not as such** (τό γε μὴ ἐόν),
>   as this cannot be brought about,
> Nor could you indicate it.]⁹

Note that, famously and amazingly, the goddess elides the subject,[10] and possibly also the predicate, when using the third-person singular forms of the verb "to be" at lines 2.3 and 2.5, while the use of the infinitive forms lack direct objects. The phrasing, in other words, is simply that ". . . is . . . ," implying at least a missing subject to the left of the ". . . is . . ." and perhaps also a missing predicate to its right. The locution is as strange in ancient Greek as it is in English translation. This seemingly cryptic way of speaking indicates Parmenides's intention to have the goddess highlight the substantive content indicated by the grammar of speech that is most easily missed, that is, that notion (whatever it may be) that binds together the terms of a sentence and that "lurks in the shadows" behind the sentence's more evident terms like nouns, adjectives, and predicative expressions. At issue here, in other words, is that which is between or among terms like "Socrates" and "snub-nosed" in sentences like "Socrates is" and "Socrates is snub-nosed."

The forms of the verb "to be" indicate being and are basic verbal elements of speech, but we should remember that their semantic functions are not relegated to explicit instances. Even sentences phrased with verbs other than forms of "to be" imply reference to being; for instance, "Theaetetus flies" entails "Theaetetus *is* flying," and/or "Theaetetus *is* the kind of thing that flies," and other context-dependent construals tacitly entailing "to be." In both ancient Greek and English, speakers can make meaning apparent by eliding explicit uses of the verb "to be"; for example, "You good," clearly stands in for, "You are good," with the meaning clear even without the "are." What is on display in the naked ". . . is . . . ," thus, is something easily missed due to its sheer ubiquity in speech, which exhibits meaning even when it is implied or elided.

Importantly, Parmenides has the goddess emphasize meaning here by connecting the ". . . is . . ." to what I translate imprecisely as "knowing" (e.g., νοῆσαι, 2.2). Subsequent chapters by Paul DiRado and Wiitala will entail unpacking the meaning of this notion further; crucial to our interpretation to follow is seeing the sense in which being here does not concern (mere) "independent existence," but instead a way in which being is bound to the meaning implied in an act of knowing, and hence is located within a world of things that are *with respect to* other things. As I will argue below, "being" here is only coherent when it implies the structure of "being in such a way as to bear meaning to a knower."

We should also note that the path of inquiry via ". . . is . . ." at 2.3 is presented as conceptually linked to the path of inquiry via "is not" at 2.5. The goddess states that, because the opposite of this path (whatever it might be) simply cannot be, we are turned back to that which has no opposite and therefore is most ubiquitous. That these paths are shown from the start to be connected and mutually illuminative will be central in what follows, as I will argue that the most otherwise-promising interpretations of the ". . . is . . ." on offer do not yield sufficient explanation of how this sense of being at 2.3 must follow from insight into so-called "non-being" at 2.5. The goddess summarizes this motivating notion in this putative "second" path through her indication of the unknowability of what I have translated as "what-is-not *as such*" (τό γε μὴ ἐόν, 2.7), with the "*as such*" intended to capture the force of the intensifier γε.[11] The task, thus, is to explain the meaning of this ". . . is . . ." and particularly with reference to its givenness to knowing and its alleged opposite, so-called non-being, though I will ultimately submit that the goddess's account entails that this notion in fact has no opposite and instead is unopposed.

Competing Interpretations of the ". . . Is . . ."

This brings us in our journey to a brief review of the major interpretations advanced, in the shadow of the thinking of being represented by substance ontology and Mill's divisions, in the last six decades or so. These are the "existential" view that Parmenides is, here, identifying an entity that exists or existence itself; the "predicative" view (or set of views) that Parmenides is after some aspect of what thinghood or identity entails, the "fused" view that Parmenides's goddess refers either to existence or predication depending on context, and the "modal" view that Parmenides's goddess identifies the modality of necessary being (and not "being" itself). Focusing on the entailments for our understanding of line 2.3, we will take each in turn, moving roughly in the order in which they held sway in twentieth-century Parmenides interpretations.

In taking up such a discussion, I am all too aware of the colossal hubris entailed when criticizing such a rich interpretive tradition and claiming to have accomplished something "new" concerning an issue about which great scholars of recent generations have thought and said so much of tremendous value. In what follows, I hope that my own reading shows itself to be deeply indebted to each of the four major lines of interpretation—"existential," "predicative," "fused," and "modal"—that have been advanced, thereby offered with pious reverence to all the previous scholars on whose work we are necessarily dependent as we continue to develop our understanding of being and contribute more to this story.

Existential

Although the view is quite old, scholars like G. E. L. Owen (1960) and Leonardo Tarán (1965: esp. 33–40) defended influential accounts of Parmenidean being with reference to the notion of existence in the 1960s.[12] According to such existential interpretations, we should count Parmenides alongside the substance ontologist and those following Mill who take being to concern the existence of entities most primarily.[13] Accordingly, line 2.3 should supposedly read something like: "The one, that [x] exists and that it is not possible for [x] not to exist." On different versions of this reading, Parmenides's point concerns either (i) the existential nature of *something* that exists or (ii) *existence itself* and, to speak convolutedly, the existence of existence. Upon (i), the subject has been suppressed, meaning something like "[something] exists";[14] upon (ii), the subject is roughly "that which has being" (τό ἐόν), meaning something like "existence exists."[15] In either case, this reading entails that 2.3 is an absolute construction, or a one-term use of "is" that, while missing an elided subject, could not be completed by a second predicative term, since the claim that ". . . exists" implies a missing subject but could not imply a missing predicate.

There are some benefits to this kind of reading. It captures our intuitive sense, which I think is right, that Parmenides is talking about something that is simultaneously "massive" and very basic. "Existence" is both that which many contemporary philosophers take to be the ultimate object of the deepest metaphysical scrutiny and that which is, on the standard modern view anyway, most ubiquitous and shared among everything in space and time (perhaps along with nonspatiotemporal entities as well). There is also a prephilosophical layperson sense in which "existence" seems to stand in for being as a whole, for it seems to be the case that any further specified sense of being—for instance, that which is bound by the copula—presupposes "existence" for further specification on the assumption that a thing must "first" be something rather than nothing before it could be any specific kind of thing.[16] The existential reader thus compellingly argues that Parmenides's poem is about being in the broadest and simplest sense.

Unfortunately, we cannot conclude our journey into the meaning of being upon such a conclusion.[17] This is because, if the existential reading is correct, then Parmenides is guilty of very basic and amateurish modal fallacies, and we are left with a powerful poetic articulation supported by extremely weak philosophical argumentation. For if his ". . . is . . ." means either "*x* exists" or "existence exists," then Parmenides has the goddess move fallaciously from *existence* (ἡ . . . ἔστιν, here meaning "it exists") into *necessary existence* entailed by ὡς οὐκ ἔστι μὴ εἶναι ("is not possible [for] [. . .] not to "be," here meaning "exist") all in the same line.[18] Of course, the claim that most (if not all) things that exist *necessarily had to* exist is nonsense; for instance, Milo the cat is here but easily could have never "existed" had his cat parents never crossed paths. At best, such a claim would require convoluted defense, which Parmenides does not have the goddess offer.

In fact, if the goddess is speaking of existence, then nearly everything she says about being is fallacious or false.[19] Some further instances of fallacy are evident when we consider the issues of knowing and nonbeing flagged above. We can inquire into and speak of that which "does not exist" like square circles and the largest prime number fairly straightforwardly, using math and logic to show the impossibility of such entities, while the goddess prohibits inquiry into and speaking of that which "is not" throughout the poem. Similarly, it would be fallacious to say that what "does not exist" *necessarily* does not exist. Milo's kittens "do not exist" because he was neutered before he could have any, but they likely would have "existed" had he not been neutered. Reading the goddess's nonbeing as "non-existence" hence is not especially satisfying philosophically; the further interpretations will show us that we should instead think of senses in which the goddess is discussing *nothingness*, or that which is "in no way," which is not coextensive with that which does not exist.[20] For instance, centaurs do not exist but are not nothing, since we can speak truly about them: compare the claims, "A centaur is half human and half horse" and, "A centaur is half human and half giraffe." Since the former is true and the latter is false, centaurs admit of true and false discourse and hence cannot be absolutely nothing; accordingly, we should carefully distinguish between the "non-existent" and, to borrow a formulation, "that x which is not F, for all values of F."

Of course, we might ultimately have to conclude that Parmenidean philosophy does entail fallacy or false assertions. But since there are more charitable interpretations that do not entail such straightforward invalidating assignments, we must consider them as viable alternatives. I do, however, suggest that we retain the existential reader's hypothesis that Fragment 2 concerns the meaning of being in its deepest and most profound sense(s) as we continue our journey.

PREDICATIVE

The most promising and recently influential of the alternative approaches to understanding Parmenidean being is the set of "predicative" readings. Such readings increasingly have held sway since the time of Alexander Mourelatos's version, first published in 1970, and we here briefly consider four such interpretations. Mourelatos initially called his reading (1) "speculative predication";[21] on his interpretation, line 2.3 reads: "X is [really and truly] Y,"[22] where "Y" discloses the true nature and structure that underlies the merely apparent nature of X (esp. Mourelatos 2008, 56–62).[23] This concerns what it is to be a *thing* with an essence, definition, or nature: the ". . . is . . ." points to the necessity of understanding beings with respect to their essential nature that "lurks beyond" mere appearance, and the sense in which the objects in space and time present themselves only partially to us while the truth of their ultimate nature remains obscure. For example, Mourelatos takes this to indicate of the other Presocratic accounts we are familiar with, like Thales's apparent view that (the) all is "*really*" water.[24]

There have been several other promising and influential versions of the predicative interpretation. These include: (2) "veridical predication," Charles Kahn's view (which changed over time) of the verb εἶναι as rooted fundamentally in a predicative sense with strong veridical and weakly existential force.[25] For example, Kahn points out that "X is Y" entails that X *truly* is Y, and also that X "exists" in the sense of being something (i.e.,

Y) rather than nothing.[26] Two further developments of such a view are (3) that of Richard J. Ketchum (1990), who argued that Parmenides means to show in 2.3 that "[what can be spoken and thought of] is [something or other]"; and similarly, (4) the 'predicational monism' of Patricia Curd (2004 [1998]) that entails reading 2.3 to imply that "*X* is [*fundamentally*] *Y*," suggesting that all things most essentially are exactly one thing, that is, that identity entails singularity and unity.[27]

Such a kind of reading represents a leap forward in our understanding of being, insofar as it helps us to break the habit of thinking of being in terms of "independent existence." Instead of drawing upon the bad habit of understanding being in the one-term sense, being here is understood as entailing complexity, since we can only understand the ". . . is . . ." once we insist upon a two-term structure for all invocations of being. Furthermore, this kind of reading strongly coheres with and seems to capture the essence of earlier Presocratic thinking concerning the true nature of reality, from at least the time of Thales's apparent declaration that all is really water, and saves Parmenides from the fallacies with which the existential reading saddles him.

But there are at least some problems with such a reading. Particularly for Mourelatos and Curd, the *Y* term in the phrase "*X* is *Y*" must be of a special sort, that is, not just any predicate, but one that fundamentally explains the *X* in its real being and hence that has some kind of basic, explanatory power. This has the strange entailment that the goddess is not speaking of *being as such*, but instead a certain *kind* of being, that is, whatever kind might exhaust the nature of a particular entity. Such a view of the ". . . is . . ." is narrowly focused on special instances of being, and we have lost the sense of the ubiquity of being for which I advocate. Replacing this is the sense of necessity that entails that a thing "necessarily is what it is," as in the example that Milo *is necessarily* a cat. While these problems are not strictly decisive, I hope we will see below the value of broadening the scope of our understanding to include other senses of being and reconsidering the function of necessity.

Perhaps more troublingly, the predicative reading saddles Parmenides with the problematic "naive metaphysical view" wherein objects are understood as discrete entities and not co-constitutive.[28] This is because the terms that could sit in the *Y* position (however many they might prove to be) must themselves be ultimately explanatory and not explicated by further terms, as in the counterexample to Thales's apparent proposition that water is two parts hydrogen and one part oxygen. From this it follows that, if such a predicative reader is correct, then we must count Parmenides among those who understood being as entailing "independent existence!"

Finally, such a reading does not quite exhaust the goddess's point. The relationship between such a sense of being and knowing is not immediately apparent, since bearing essential properties does not clearly connect to these properties being known.[29] Furthermore, such a reading forces a somewhat awkward rendering of nonbeing, the "negative" route, and what-is-not *as such* at issue from lines 2.5–8. For the predicative reader, particularly as represented by Mourelatos and Curd, the goddess seeks to show that nonbeing is not especially helpful as a means of inquiry into the true nature of something. For example, to say that X is a "non-cat" does not offer much insight into the nature of X, as a "non-cat" could be a dog, a pizza, the number 5, a fire hydrant, the Platonic form "justice," and so forth. While it is certainly true that the term "non-cat" is not especially disclosive of truth and hence not helpful in our quest to understand being, such an understanding of "nonbeing" does not seem to capture the intensity of the goddess's denunciations of "what-is-not" throughout the poem: it is difficult to imagine, for instance, her aggressive attack on mortal thinking for taking being and non-being to be "the same and not the same" (6.8–9) referring to confusion between being a cat and being a non-cat, or unpacking her declaration that "never shall this prevail, things that are, are not, and do restrain your thought from this route of inquiry" (7.1–2) with reference to such a view of nonbeing. For this reason, the predicative reading does not capture the force with which the goddess speaks against nonbeing throughout the poem, nor entirely exhaust the insight into being attained through an encounter with nonbeing that the poem enacts. Hence, our quest for insight into being continues, albeit with many powerful resources from such a reading now at our disposal.

Fused

One way of understanding Parmenidean being—and one that is roughly as old[30] as the more influential versions of the existential and predicative readings discussed above, but that was developed more thoroughly in recent decades[31]—is to take the goddess's articulations to entail a fused sense of the ". . . is . . ." that can indicate *either* existential *or* predicative force (including that of identity, veridicality, location, or any combination of these) depending on context. On the fused interpretation, 2.3 reads: "X is [Y] and it is not possible for X not to be [Y]," where Y can be supplied or not based on context. This interpretation follows from the assumption that existence and predication are ultimately separate but nevertheless frequently co-present.[32]

This seems to be a promising way of resolving the apparent tension between the existential and predicative senses of "to be." Fused readers are rightly unwilling to reduce Parmenidean being to any one of the senses of being recognized in the kinds of divisions proposed by Mill. For the fused reader, we can pick whichever reading makes more sense in a given context and not fear a conflict.

Certainly there are pragmatic gains, but there are three reasons why I propose that we continue our quest into being without resolving the issue as the fused reader suggests. First, such an orientation still entails assuming the fundamental distinction between the alleged senses of existence and predication in the first place, tending to emphasize (if not necessarily deferring to) the existential sense in most cases.[33] In other words, the interpretation entails presupposing the kinds of divisions into being made by people like Mill; I suggest that we instead heed the goddess's declarations in Fragment 8 that being is "whole," "one," and "undivided."

Second, such a reading threatens to be question-begging in allowing the interpreter to choose whichever interpretation serves a given point while maintaining that the choices are somehow fundamentally separate; for example, the question as to whether the ". . . is . . ." of 2.3 has an elided predicate threatens to be answered arbitrarily. Better, I think, is to understand the ". . . is . . ." as intending some central thing that is whole, one, and undivided, and around which its various expressions revolve, even if (and here I agree with the fused reader) we need not understand it as strictly univocal.

Third and finally, the fused interpretation entails treating the ". . . is . . ." as a merely *linguistic* device. Thus conceived, the ". . . is . . ." does not have a clear ontological correlate, since it must indicate *either* the semantically "absolute" (or syntactically "complete") existential "is" *or* a semantically incomplete predicative sense, or the two simultaneously in a way that still preserves their alleged mutual disjunction. Thus, if the "fused" readers are correct, the goddess is talking to us about our human language and not being itself! Given these problems, we should press onward in our journey.

Modal

Particularly via John Palmer's defense (2009: esp. 45–50 and 83–105), a new reading of the goddess's "... is ..." has gained some traction.[34] Upon this modal reading, the goddess says at 2.3 that "X is *necessarily* [X or Y]," and hence must be understood as referring to that which is (as it is) necessarily, as a circle is necessarily round or two and two necessarily are four. On this view, the goddess is not talking about being itself at all, but merely a certain sense of being: that which is necessarily (2.3) and that which is not necessarily (χρεών ... μή, 2.5).[35] The goddess's point, in other words, concerns our recognizing the necessity-contingency distinction.

On this reading (and borrowing the version developed by Jeremy DeLong in this volume), the goddess is speaking to a sense of necessity that mortals have missed, which is a point about the necessary identity of divine (i.e., eternally self-same) entities that mortals have mistakenly treated on the model of contingency (e.g., anthropomorphically). The modal reader holds further that turn enacted in Fragment 6 and its consideration of the intermixing of being and nonbeing concerns contingent being.[36]

There is value to this kind of reading, particularly since it encourages us to reconsider the sense in which the goddess is interested in what *must* be the case and thereby address what I identify as a weakness in the predicative interpretation. For predicative readers, the "... is ..." captures what the thing *necessarily* is,[37] for example, that Milo is (really and necessarily) a cat and as opposed to whatever further nonessential predicates Milo might have like affectionateness, blue-eyedness, and so forth. This yields the unsatisfactory results that the goddess is speaking of being in a highly qualified sense, that essential predicates could not be mutually co-constitutive, and hence that things are ultimately "independent"; the modal reading is thus valuable for encouraging us to rethink the role of necessity in the account.

But suffice it here to point out that the modal reading is unfavorable insofar as it leads to awkward renderings for several of the goddess's points. Considering knowing and nonbeing will again be illustrative. The modal reading suggests that we cannot know, indicate, or point out that which necessarily is not; but this is unpersuasive, for clearly it would be false to suggest that we cannot know, indicate, or point out that two and two necessarily does not equal five, or that I necessarily am not taller than myself. In this case again, the interpretation has not made adequate sense of the roles of knowing and the insight into the impossibility of nonbeing in allowing the goddess to indicate the sense of being that she has in mind.

Furthermore, the goddess seems to refer elsewhere in the poem to the routes identified in 2.3 and 2.5 without explicitly modal language. At line 7.1, the goddess merely identifies the route of nonbeing as the prevalence "to be of the things that are not" (i.e., without reference to their *necessarily* being or not being); at lines 8.1–2, the goddess speaks of the "single story of the route still left, that '. . . is . . .'" (again without reference to necessary being).[38] Hence I propose that we arm ourselves with the notion of necessity while not reducing the goddess's points to the issue of modality as we proceed forward in our quest.

Being-as-such

Having traveled along this storied route, we are prepared to consider a strongly predicative hybrid reading, which I will call "being-as-such." The aim is to avoid the suppositions of later senses of "divided being" privileging so-called existence like those following substance ontology and Mill, while drawing upon the strengths of the four kinds of readings considered above. Hence, I propose a way, admittedly convoluted in its phrasing here, of grasping the meaning of 2.3: "that a thing is and, by virtue of the '. . . is . . . ,' is picked out as a being with a structure (e.g., that admits of true and false assertions concerning its predication)," where the ". . . is . . ." is understood as entailing a corresponding ontological principle (instead of mere linguistic device) through which entities are as they are. Drawing upon predicate logic notation, we would represent the construction in 2.3 without the universal ($\forall x$) or existential ($\exists x$) operators, simply as Φx, or 'a [variable] thing [x] is in a certain way [Φ].'[39] While there is of course weak existential force in such an assertion (i.e., in a way that is not especially interesting philosophically), the emphasis is not on existence or universality, but instead on the binding relationship between x and Φ. The ". . . is . . . ," in other words, is implicit in the structure of the articulation Φx between its two terms. If we were to represent this bindingness with the B predicate variable, we could then formulate the further, philosophically consistent proposition $(\forall x)Bx$: "all things are bound things."

On the being-as-such reading, the ". . . is . . ." concerns ontological grounding, and the necessary sense in which that being indicated is structurally complex (following the predicative reading), albeit through a principle that is itself ontologically simple (following the existential reading, but without importing the modern notion of existence). Such a structure is a necessary entailment of being that captures all of, but does not reduce to any one of, the divided senses of being following Mill.[40]

Our journey indicates that we are best to turn our thinking away from conceiving being as "independent existence." By connecting being to knowing, Parmenides has the goddess show that the sense of being at issue here is not a principle of self-relation, but instead a showing forth of the self for another in such a manner that can be known by another. This is one sense in which the object of the goddess's discussion is something that is experiential: to be is to be a self with a predicative structure that is present to another. I do insist upon emphasizing that we need not understand this anthropomorphically: for example, a cat is a cat in accord with other animals and entities in its environment, and hence it *is* a cat with respect to its world. This reading furthermore could find support through a defense of a kind of panpsychism, but I will set aside that issue here. In any case, the sense of being that emerges from the ". . . is . . ." is one that points to the situatedness of being within a web of relations, as opposed to naive separation, with knowing as a paradigmatic mode in which being is for others and not merely for itself.

Similarly, we can grasp this sense upon experiencing the impossibility of nonbeing, for this sense of being must contrast with an unknowable quasi-opposite that is impossible, namely, the so-called "what is not '*as such*'" (2.7). While taking this to mean "what does not exist" yielded logical fallacies, understanding it as an impossible, unknowable, ungraspable nonentity possessed of no predicative structure is far more appealing, for an understanding of the necessary being of any X that is posited is encountered through the revelation of the impossibility of nonbeing as such. In short: when we try to think of nothingness, "pure" nonbeing, what-is-not *as such*, or that "X that is not F, for all values of F," we draw a blank. More precisely, we *cannot* draw a blank, since thinking always entails content. This encounter with the impossibility of what-is-not *as such* in thought is a first step toward our realization of being, which must *be* in all instances and simply cannot not be in being as is also the case in speech. This sense of being is itself simple, but it demands complexity in all instances, since all being is being in a certain way. Put differently, any x will always be Φ, though the value of Φ will vary. Once the thinking mind grasps the impossibility of nonbeing *as such*, it must take up that ubiquitous (or "necessary") sense in which being is indicated through any and all declarative indications either in thought or speech. If we are able to understand such a notion as such, we seem to be left with insight into the necessity of determinately intelligible predicative structure, including essential and inessential predication, in all instances of being.

Hence, one of the goddess's lessons is that "to be" entails "being in a particular manner" (cf. Ketchum 1990); but more deeply, this structure is the most unyieldingly ubiquitous aspect of all the beings. In this way, the modal reading offers some insight, since it entails emphasizing the sense in which being *necessarily must be and cannot not be* (οὐκ ἔστι μὴ εἶναι, 2.3);[41] but rather than interpreting this sense of logical necessity, I submit instead that we interpret this as a sense of ubiquity, and the necessary ubiquity of being in all instances.

Like the existential reading, this is an ontological interpretation of the ". . . is . . . ," and it concerns ontology at its most basic level. But as I have stressed,[42] we need not understand ontology as fundamentally rooted in the modern notion of "existence" for which there is no clear correlate in the Greek. In accord with the other predicative readers, we can understand this sense of the ". . . is . . ." as a continuation of the Presocratic project: it captures the spirit of ultimate truth as disclosed through an account of being as being-of-a-certain-nature from the tradition of Presocratic ontology but with the benefit of not requiring that we understand the whole of reality on the model of a material object in space and time (e.g., water), indicating Parmenides's greatest achievement beyond his peers. Drawing inspiration from the fused reading, we have the resources to undermine the fundamental distinction between existence and predication without reducing the goddess's ". . . is . . ." to a mere linguistic device. And finally, following the modal interpretation, this reading entails at least some modal force concerning necessity,[43] since being *in its ubiquity* (or in its *necessity*) is at issue.

While the reading I am endorsing is indeed predicative, I emphasize two distinctions from previous versions. First, Parmenidean being thus conceived is not relegated to essential, definitional, or identity-marking senses, but instead about being in *all* instances, that is, whether concerning essential or inessential properties. In this way, the goddess's ultimate target is *being*, and not strictly accounting for the nature of being an entity. Second, the goddess's ". . . is . . ." on this reading concerns not naively individuated thinghood but instead a *binding* aspect of our collective being. *All* beings share in being-as-such, and this truth about our collectively shared nature constitutes the sense in which we could truthfully call Parmenides a monist. This form of monism is also consistent with some kind of pluralism, since all entities could indeed share in one thing (i.e., being) while being different in other senses. In this way, Parmenides offers us a version of the claim "all is one" that does not preclude the subsequent claim "all is not many."

Regardless of how we might ultimately account for it, the Parmenidean insight into being proves very hard for us mortals to grasp. The philosophical tradition thereafter and the quest we have considered attest to that difficulty. Similarly, the rest of the poem after Fragment 2 depicts the goddess helping us by reminding us of the difficulty of such insights (esp. Fragment 6) and giving signs pointing toward the senses in which this abstract notion of being is accessible to mortals only in mediated ways via "that which has being" (τό ἐόν, esp. Fragment 8; cf. Sanday's chapter).

Hence, even if my account has offered at least some insight into being, we must press onward in our journey to grasp the meaning of being. In the subsequent two chapters, DiRado and Wiitala will call our thinking to the provocative claim that being and knowing are (in some sense) "the same" to bring further insight into the meaning of Parmenidean being.

Notes

1. That is, using ellipses to represent these as unresolved copulas and infinitives with elided but implied objects. For the grammar of such a rendering, see, e.g., Mourelatos 2008, 55 and Bredlow 2011.

2. Overviews of the history of this reception include Mansfeld 1964, 51–55; Tarán 1965, 33–40; Mourelatos 2008, 269–76; and Curd 2004, 34–51. Bryan 2020, esp. 230–35 rightly illustrates that the Parmenidean insight into being is far greater than that of mere "clarification," i.e., concerning the "meaning" of being in the modern analytic-philosophical sense of "meaning." Hence, my goal in this chapter is to speak a bit and as best I can about what it is that we *experience* when sorting through Fragment 2's provocations. A complementary account of this notion as experiential is Robbiano 2011.

3. Aristotle writes in *Metaphysics* 4.2 that being is said in many ways, but that all ways intend a central way; unfortunately, the surviving manuscript tradition does not allow us to make clear what exactly Aristotle thinks that central way is (see Menn 2021). On my view, there is good reason to reject the notion that Aristotle ultimately was a substance ontologist himself and instead follow those arguing that we are better to understand Aristotle's view of the central sense of

being as "activity" (ἐνέργεια), particularly following evidence in *Metaphysics* 7–9; see Kosman 2013 and Gonzalez 2019.

4. A concise overview is Frede 1993, 44–46.

5. On the history of this debate, see Mourelatos 2008, 351–52.

6. ὑπάρχειν has become associated with the verb "to exist," but see Kahn 2012, 39 and 54–61 on why this term initially meant something like "emerging from the shadows" and only later became associated with the modern sense of existence under the influence of philosophical theology.

7. An emendation for ἀληθείη in the manuscripts.

8. Variant reading of παναπειθέα.

9. Translation reproduced from Smith 2020, 281.

10. Concerning the scholarly consensus of this point and its implications for our understanding of the ". . . is . . . ," see Mourelatos 2008, 47.

11. Cf. Miller 2006, 3–4. Reading γε as "*as such*" is controversial, and the term might have other senses, but the other meanings would not invalidate the account of being we are developing. In short, the phrase is ambiguous, but while I think all its many construals would support the reading for which I advocate here, the matter is clearest when we read γε as an intensifier.

12. Tarán also describes the history of this view and its development in the nineteenth and early to mid-twentieth centuries. Gallop 1984, 7–12 is a concise explanation and defense of the existential reading.

13. Curd (2004, 34 n. 29) rightly points out that existential readers tend to base their arguments on pretextual understanding of the meaning of being, whereas the predicative readers rely more heavily on other ancient Greek textual sources for their interpretation of Parmenides. This is relevant to the history of understanding being I describe above.

14. See, e.g., Cornford 1939, 30–32 or Owen 1960, 95 (who argues that the subject of "exists" emerges dialectically in the text). As Mourelatos (2008, 271) points out, other proposals include "reality," "truth," "the One," "the Route," and "what can be thought or spoken of"; see Mourelatos 2008, 270–72 for critical discussion.

15. This is the position of Tarán (1965, 38), who translates the ". . . is . . ." as "exists" and writes, "The point of departure of Parmenides is that there is existence, whatever the existent may be."

16. I have in mind conceptual, and not temporal, priority: on this view, having any property conceptually presupposes existence, even if there will never be anything that temporally exists before having any properties.

17. For more developed argumentation against importing the notion of existence into ancient Greek thinking, see the work of Charles Kahn and Lesley Brown. Kahn 1966, 248 represents an early, programmatic articulation of the point that "the Greeks did not have our notion of existence"; Kahn 2003 and 2012 are more thorough and detailed studies. While Kahn's account changed over time, he

argued persistently against reading the ancient Greek εἶναι with strong existential force. Brown 1986 focuses on being in Plato's *Sophist* but contains insights for our broader understanding in ancient Greek philosophy, further developed in Brown 1994.

18. See especially Lewis 2009 for development of this problem. Since the "existential" interpretation would render most or all of Parmenides's goddess's claims invalid (cf. Ketchum 1990, 186 and Bredlow 2011, 284), the interpretation seems dubious.

19. Ketchum (1990, 186) rightly states, "Most of what Parmenides tells us about why what-is-not is unthinkable, etc., is false and, if I may say so, fairly obviously false on the existence interpretation"; cf. Mourelatos 2008, 273–74 and Palmer 2009, 74–82.

20. For defense of this point, see Owen 1971, 244–48.

21. See Mourelatos 2008, 58–60 for explanation of the choice of this term, which Mourelatos later sought to modify (as summarized at Curd 2004, 40n46). Bredlow (2011, 292) similarly accounts for the ". . . is . . ." as the "is of whatness or definitional predication."

22. For Mourelatos, this entails reading the ὅπως (2.3) and ὡς (2.5) adverbially, i.e., indicating *how* it is the case that ". . . is . . ." and "is not" (Mourelatos 2008, 49–51). Mourelatos shows good reason to interpret these adverbially, but since (as he himself admits) nothing ultimately hinges on such a reading definitively, I defer to the more common "that" in my translation.

23. Mourelatos's view has evolved somewhat over time: see Curd 2004, 39–41.

24. Of course, we have no record of anything Thales said, let alone anything to substantiate such a view definitively. Less speculatively, Mourelatos finds such a use of the Greek verb grammatically and conceptually in Xenophanes and Heraclitus: see Mourelatos 2008, xxi–xxiii.

25. Kahn 2003 and 2012 are the most definitive and developed accounts of such a view.

26. Although Kahn's view has strengths that the existential view does not, Kahn (2002, 90) still admits that his reading must render Parmenides's goddess's arguments fallacious.

27. Curd's view is similar to but not coextensive with Mourelatos's; see Curd 2004, 40 n. 46.

28. See Mourelatos 1973, 28 and 1999, 122–24, and Bredlow 2011, 288.

29. Mourelatos (2008, esp. 63–71) makes compelling points about the connections among being, knowing, and "the quest" (διζήσιός, 2.2) for the truth (ἀληθείη, 2.4) of what lies concealed. I find his account entirely persuasive, but note that it would, I think, be equally compelling as an argument for *being itself* (for which I argue) and not the ". . . is . . ." of identity or speculative predication. Ketchum (1990, 175–76) offers insight into the fit of his reading with the elliptical Fragment 3 and the being-knowing identity thesis, though his reading commits

him to one of the two possible readings of Fragment 3 described by Wiitala in his chapter herein.

30. See Furth 1968, esp. 117–27.

31. See, e.g., Pelletier 1990 and Wedin 2014.

32. On the version of the "fused" interpretation I have in mind (represented, e.g., by Wedin 2014), this is not fallacious on Parmenides's part. Some have argued that Parmenides fallaciously conflates the two, e.g., Kirk et al. 1983.

33. Cf. Curd 2004, 34 n. 28.

34. DeLong develops Palmer's reading in his chapter in this volume.

35. These are roughly captured by the □ and □¬ of modal logic, respectively.

36. Roughly captured by the ◆ operator.

37. Bredlow 2011, 290–92 well illustrates such a reading.

38. Cf. Ketchum 1990, 167–68 for further discussion.

39. Mourelatos (2008, 51–52, drawing on Calogero 1932) defends similar but not identical points.

40. This is similar to but stronger than the fused reading, and is spoken to well by Hintikka (1980, 6), who writes that the distinctive feature of the Parmenidean view of being is "the absence of any distinction between what we twentieth-century philosophers would take to be different senses of 'is' [. . .], viz. the 'is' of existence, the 'is' of predication, and the 'is' of identity."

41. That is, necessity in accord with the □ of modal logic.

42. And following the readings of Kahn and Brown cited above.

43. Albeit not that of the □ operator. Cf. Mourelatos 2008, 72 n. 61.

Chapter 5

The Veridicality of Noein and the Particularity of Noos in Parmenides's Poem and the Continuity Between Parmenides's, Homer's, and Hesiod's Usages

Paul DiRado

In his poem, Parmenides seems to use the words noos and noein in conflicting ways. There are numerous places in which Parmenides apparently connects the pair directly with being and truth, and thus seems to imply that this activity is unerring. Yet there are also several passages in which Parmenides describes a wayward noos that falls into perplexity, partiality, and error: a noos that walks the path of Doxa, rather than Truth. This tension has been well noted in the scholarship. To take just one example, in his landmark study of noos and noein prior to Anaxagoras, Kurt von Fritz notes the dual uses of noos and noein in the poem.[1] As a result, he decides to translate noos and noein as "thinking" and "to think" in the poem, despite the fact that such a translation is inconsistent (1) with his earlier analysis of the words in other thinkers, which demonstrates that they do not pick out a strictly cognitive activity, and (2) with the primarily intuitive sense that he thinks both words possess throughout the tradition and in Parmenides. Yet even with this translation, von Fritz is not able to give an interpretation of precisely how the words function that he finds completely satisfactory.[2]

The assumption that underlies this apparent tension in Parmenides is that the words noos and noein have the same basic semantic range within the poem. Noein, on this account, is a verb that means something like "to see, grasp, or think about something," and noos is a noun that either picks out the part of the person that does this activity, and so means something like "mind," or serves as a verbal noun of noein and so means something like "thinking." The closely related word noēma would, on such an interpretation, pick out that which is grasped or thought about by either noos or noein. Adopting such a reading, the puzzle is why Parmenides sometimes appears to speak as if noos and noein are veridical, as the natural interpretation of minds and thoughts is that it is possible for them both to hit upon the truth and to err.

In this chapter, I will propose an alternative interpretation: noos and noein slightly diverge in meaning. Recognizing this divergence will help clarify both how the words work in Parmenides and also the continuity between Parmenides's usages and those of earlier authors. I will argue that the noun noos means something like "a person's individual or particular grasp of a situation." As a result, it can indicate any grasp of a situation, even mistaken, incomplete, or foolish ones. By contrast, I will argue that the verb noein means something like "to actually or correctly grasp a situation," and thus it has a fundamentally veridical sense. Noēma, "that which is grasped," can refer either to the content of someone's noos (and so can be nonveridical) or to the thing grasped in noein (and so have a necessarily veridical sense) depending on the context.

The relationship between noos and noein that I will advocate for in this chapter is somewhat analogous to the relationship between "understanding" and "to understand" in English. The primary sense of "to understand" in English is veridical: when I say that I understand something, I am claiming that I have gotten it right. When someone has not gotten it right, the most natural way of articulating the point is to say that they did not in fact understand it. To be sure, there are nonstandard uses of "to understand" where the verb loses its veridical force, and these uses are perfectly sensible in English, e.g., "they understood me incorrectly." But when philosophers restrict language in a technical way, they restrict the verb to contexts that are veridical at least in part because that usage best captures the basic sense of the verb even in normal language. By contrast, the noun "understanding" refers to how some individual grasps something, and as a result includes cases where someone's understanding is incorrect. Indeed, one of the more common uses of the noun is as a hedge in phrases like, "It is my understanding that . . . ," which emphasize that the understanding is mine and, as such, might not have perfectly hit the mark. My argument is that the situation with noein and noos is comparable, though with this comparison I do not mean to suggest that the meaning of noein overlaps with "to understand." If nothing else, it is far easier to find nonstandard, nonveridical uses of "to understand" in English than it is to find nonveridical uses of noein: I could only find one potentially nonstandard usage in all of Homer and Hesiod.

My argument will proceed by first showing that the basic semantic ranges of noos and noein slightly differ in their uses in Homer and Hesiod. There, noein possesses a fundamentally veridical sense that noos does not. I will then show that this sense is preserved in Parmenides's uses of the noun and verb. As such, there is no linguistic or terminological tension between the wayward noos that walks the pseudo-path of Doxa and the noein that is fundamentally linked to being and Truth. My hope is that noting how noos and noein differ will provide helpful resources for interpreting Parmenides and other Presocratic philosophers coming from Homer and Hesiod.

Features of Noos and Noein in Parmenides's Predecessors

Before turning to Parmenides's uses of noos and noein, I will first motivate my interpretation by examining how earlier authors used the words. There is general scholarly agreement that noos is not intrinsically veridical in Homer and Hesiod. As such, in this chapter I will focus primarily on noein. However, it will first be helpful to trace the basic way that noos functions: I still take the noun and verb largely to overlap in meaning besides this one point of divergence. Indeed, there is general agreement that the noun noos is the source of the verb noein. In the interest of space, I will rely primarily on prior studies of noos.[3]

In Homer and Hesiod, noos is generally taken to have three basic meanings: (1) It primarily functions as an action noun that picks out an individual's grasp of the meaning or significance of some situation. This grasp frequently blurs the line between what moderns would distinguish as perceptual and intellectual, and frequently has characteristics of both. (2) Derivative upon the primary function, it also gets used to indicate something like an individual's concrete plan of action (not a wish or desire), and as such is frequently used in what moderns would characterize as volitional contexts, though intellectual elements aren't completely absent from these uses. Indeed, in many cases noos seems to be perceptual, intellectual, and volitional at the same time. There are also uses that seem to involve both grasping the present situation and grasping or planning what to do in the future. (3) Finally, noos is also used to pick out an individual's general tendency or ability to grasp situations or to formulate plans of particular sorts. In this third usage, some scholars characterize it as a "psychic organ," though I agree with von Fritz that this characterization is at best misleading and that "Homeric Greeks did not make the rather abstract distinction between an intellectual organ and its function but . . . if they had made the distinction they would have considered the νόος a function rather than an organ."[4] Noos has a highly intuitive character; while a process of reasoning might be involved in the emergence of noos, the noos itself is an immediate grasp.[5] These three uses are thus highly continuous, and to express that continuity I will translate all three with "grasp": Helen's noos can refer to her grasp of the situation, her grasp of what to do given the situation, or her tendency to grasp situations in some particular way. I will render the verb noein similarly.

Determining someone's noos is often said to supply real insight into who that person is, which makes sense: if you know how I grasp situations in general, how I grasp this situation in particular, and what I plan to do on the basis of my grasp, you know an enormous amount about me and how to get the better of me in an exchange. The fact that two people have a different noos does not necessarily mean that either's noos is wrong; two people can both have a correct grasp of the situation, but what that situation means to them might vary greatly based on how they are situated within it and their character. This does suggest that an individual's noos will tend to be partial, however, and there is a risk that individuals might only grasp that part of the situation that immediately pertains to them and so fail to have a complete picture. Indeed, it also opens the possibility that someone will have a completely incorrect, confused, or childish grasp of the situation, and that foolish things will be done ἀιδρείῃσι νόοιο, in the ignorance of someone's noos (Hes. WD.685).[6]

Passages Where Noos and Noein Are Juxtaposed

The first passages in Homer and Hesiod to discuss are those where noos and noein are used together in such a way as to show how the words relate to one another. I found four such passages, all in Homer. Emphasized in these four passages is either the veridicality of noein in contrast to someone's noos or the particularity of someone's noos in contrast to the universality of noein.

The first such passage is at *Iliad* 9.104–5, where Nestor sets out to offer Agamemnon counsel as to how to reverse the Greek's flagging war prospects. He begins his speech by establishing his credentials to offer council:

> οὐ γάρ τις νόον ἄλλος ἀμείνονα τοῦδε νοήσει
> οἷον ἐγὼ νοέω ἠμὲν πάλαι ἠδ' ἔτι καὶ νῦν
>
> [Nor shall someone else grasp (noēsei) a better noos than this,
> since I grasp (noeō) both past and still now.]⁷

Nestor begins by praising his noos, which in this context means his grasp of what counsel to offer Agamemnon given the situations the Greeks find themselves in. No other person can offer Agamemnon better counsel, suggesting that in principle Nestor's noos is superior to the noos of others. The cause of the superiority of Nestor's noos is that he grasps (noeō) how the situation in the past has led to the present situation: namely, that the fortunes of the Greeks turned when Agamemnon took Briseis from Achilles (*Il*.9.106–7). He grasps all of this (in the present tense) and thus holds together both past and present. As a result, he can offer the best possible counsel. By contrast, those with inferior noos presumably fail to grasp adequately or fully—as I will argue below, they fail to noein—how the present situation is related to the past situation. Those with inferior noos might not have an outright false grasp of the present situation, but their individual grasp is at best partial or incomplete, and so their noos fails to actually grasp the situation that the Greeks are currently in. Their counsel would thus be suspect.

A similar juxtaposition of noos and noein is at *Odyssey* 19.478–79. Eurycleia has figured out that the visiting stranger is Odysseus, and attempts to reveal this to Penelope through a glance.

> ἡ δ' οὔτ' ἀθρῆσαι δύνατ' ἀντίη οὔτε νοῆσαι:
> τῇ γὰρ Ἀθηναίη νόον ἔτραπεν
>
> [But she [Penelope] could not meet her glance nor grasp its significance (noēsai):
> for Athena turned around her noos.]⁸

Once again, we see someone who has a noos—a grasp of the present situation—that is either false or, at the very least, distorted or incomplete. And again, the inferiority of her noos is connected with her failure to noein, a failure to grasp correctly the meaning of Eurycleia's sign and thus the entire situation that would have been decoded for her through that sign.

At *Iliad* 10.224–26, the juxtaposition of noos and noein does not emphasize the veridicality of noein in comparison to noos, but rather the universality of noein in comparison with the particularity of noos: how an individual's noos varies based on the particularities of the situation and how an individual's grasp is embedded within the larger situation. In this scene, Diomedes requests that someone join him on a scouting mission to the Trojan camp. He explains:

σύν τε δύ' ἐρχομένω καί τε πρὸ ὃ τοῦ ἐνόησεν
ὅππως κέρδος ἔῃ: μοῦνος δ' εἴ πέρ τε νοήσῃ
ἀλλά τέ οἱ βράσσων τε νόος, λεπτὴ δέ τε μῆτις.

[With two going out together one grasps (enoēsen) before
    the other
where advantage lies: a man alone, even if he does grasp
    (noēsē) this,
all the same his noos is late (or slow), and his plan is weak.]

According to this passage, it is possible for two people working together and one person working alone both to grasp the situation such that they detect where advantage or profit can be found. But attaining this grasp takes longer for the person who goes alone. The cost of this is that the noos of the lone person—in this context, the grasp of the situation, where advantage lies, and how to go about exploiting it—comes so late that the opportunity to take advantage of the situation might have been missed. Plus, being alone hinders the person's ability to exploit the opportunity: an individual can accomplish less than two can. What is emphasized here is not the veridicality of noein: the noos of the person who goes alone is not described as incorrect, or as failing to be a grasp of where the advantage lies. Nevertheless, the noos of the lone person differs from the noos that would be had with another. The lone person may well discover that the advantage can no longer be exploited, or that it cannot be exploited by oneself. The implication of this passage is that different people (or even the same person in different settings) will have a different noos even when they are able to noein the same thing, as individuals are embedded differently within the shared situation. These differences shape their individual situations: a mighty warrior and a weak coward can both correctly grasp the situation on some battlefield—say, that one side is winning—but how they individually relate to that situation will vary widely based on their individual characters and capacities for reacting to it. Thus, their individual noos will vary even though both are able to noein the same situation on the battlefield.

Conversely, we also find that a person with a good noos is able to noein truths that others will be incapable of grasping at *Odyssey* 20.364–70. Here, Theoclymenus, a prophet, responds to the taunts of the suitor Eurymachus that Theoclymenus should be guided out of the feasting hall:

'Εὐρύμαχ', οὔ τί σ' ἄνωγα ἐμοὶ πομπῆας ὀπάζειν:
εἰσί μοι ὀφθαλμοί τε καὶ οὔατα καὶ πόδες ἄμφω
καὶ νόος ἐν στήθεσσι τετυγμένος οὐδὲν ἀεικής.
τοῖς ἔξειμι θύραζε, ἐπεὶ νοέω κακὸν ὔμμιν
ἐρχόμενον

[Eurymachus, in no way do I bid you to give me guides:
I have eyes and ears and my two feet
and a noos in my breast that has in no way been made shamefully.
With these I go out the door, for I grasp (noeō) upon you evil coming.]

Theoclymenus does not require or want a guide from the suitors. He possesses a well-formed noos—in this context, a general tendency to grasp situations well or with insight[9]—with which he can guide himself. Implicit in his statement here is that the leering suitors do not possess such a noos; perhaps their noos is not well formed, or perhaps they do not have any grasp of the situation at all. As a result, they do not grasp the evil that is coming upon them, while Theoclymenus does and can respond appropriately. Thus, the suitors are the ones who actually need a guide, that is, Theoclymenus the prophet who has grasped their doom.

Other Passages in Homer and Hesiod that Use Noein

The four passages just examined all show that noos is highly particularized in ways that can involve error or an incomplete grasp, and that noein means to grasp things correctly. Examining all the verb's uses in Homer and Hesiod will demonstrate that these passages are representative.

In the Homeric and Hesiodic corpus, the primary use of noein and its derivatives occurs in the context of someone grasping the immediate perceptual presence of something and then reacting in some appropriate way, that is, in some way fitting given their character and the situation. In the *Iliad*, this almost always occurs on the battlefield. People see their foes or the turning tide of battle, either as a result of actively seeking them out or by happenstance, and react. These uses are always veridical: no one is ever said to grasp something that is not actually occurring. Indeed, these uses frequently involve picking out the person or event that is most relevant from the chaos of the battlefield. A representative passage is at *Iliad* 5.589–90:

> τοὺς δ' Ἕκτωρ ἐνόησε κατὰ στίχας, ὦρτο δ' ἐπ' αὐτοὺς
> κεκλήγων

> [And through the lines of battle Hector grasped (enoēse)
>   them, and sprang for them
> crying aloud.]

However, in the *Iliad* and elsewhere this usage also occurs in more mundane situations, such as *Iliad* 9.223, where Odysseus grasps (noēse) that his compatriots are ready to transition from eating dinner to attempting to persuade Achilles to return to battle, prompting him to begin a speech. The important role that noein plays in shaping how a situation subsequently unfolds is often emphasized when the text describes what would have happened had someone failed to grasp something, or what (usually bad) results would counterfactually follow if someone's presence was grasped by another. For example, at *Iliad* 20.291 we are told that Aeneas would have been slain if not for the fact that Poseidon grasped (noēse) the danger he was in.[10]

Similarly, noein is used in contexts where what is grasped is not an immediately perceptually present object or state of affairs. For example, at *Iliad* 1.342–44:

> ἦ γὰρ ὅ γ' ὀλοιῇσι φρεσὶ θύει,
> οὐδέ τι οἶδε νοῆσαι ἅμα πρόσσω καὶ ὀπίσσω,
> ὅππως οἱ παρὰ νηυσὶ σόοι μαχέοιντο Ἀχαιοί.

[For surely he raves in his ruinous heart
and does not know to grasp (noēsai) ahead as well as behind
as to how the Achaeans shall fight in safety besides the ships.]

Here, we see Achilles accuse Agamemnon of failing to realize that someone must grasp what the situation will be in the future (and thus plan or decide how to act accordingly). The passage also tells us that it is possible to grasp what the situation was in the past. Noein also gets used in the context of grasping the possibility of something, such as when Nestor says that he has never seen nor grasped (noēsa) such wondrous horses as those possessed by Odysseus (*Il.*10.550). Further, someone can grasp that something is not taking place and how that is significant, such as when Odysseus grasps (noēse) that dogs are not barking when someone approaches and infers that it is an acquaintance rather than a stranger (*Od.*16.5). In all these contexts, noein is veridical: it picks out the activity of grasping what the situation really is, will be, what is possible, and so forth.[11]

Though rarer, there are also clearly veridical uses of noein that involve grasping a conceptual distinction of some sort. The most obvious example occurs at *Works and Days* 11–13:

οὐκ ἄρα μοῦνον ἔην Ἐρίδων γένος, ἀλλ' ἐπὶ γαῖαν
εἰσὶ δύω· τὴν μέν κεν ἐπαινέσσειε νοήσας,
ἣ δ' ἐπιμωμητή· διὰ δ' ἄνδιχα θυμὸν ἔχουσιν.

[So, after all, there was not one kind of Strife, but over the earth
there are two: the first, someone would praise upon grasping (noēsas) it,
but the other is blameworthy: their tempers are distinct.]

The individual who correctly grasps the distinction between these two kinds of strife will praise the one and avoid the other. There are two other instances where someone is said to grasp a conceptual distinction of some sort: Telemachus announces that he now grasps (noeō) things pertaining to good and bad in relation to the suitors, whereas before he was a child (repeated at *Od.*18.228 and *Od.*20.309), and Zeus is said to grasp (noēsas) all things such that it will not escape his notice what sort of justice is practiced within different cities (Hes.*WD.*268).

There are also occasions when noein is used to indicate that someone is penetrating through falsehood, uncertainty, or deception, and so when the veridical force of the verb is thoroughly emphasized. One instance of this sort occurs when Helen is initially taken in by Aphrodite's disguise as an old woman, but then grasps (enoēse) various signs that indicate she is actually in the presence of the goddess (*Il*.3.396).[12] Another similar use occurs when Priam is told to pray that Zeus send him a clear sign that he will be safe if he goes to the Greek ships: one he can clearly grasp (noēsas) with his own eyes (*Il*.24.294, then repeated nearly verbatim in Priam's prayer at *Il*.24.312).[13] The verb is also used in the context of penetrating human obfuscations or deceptions, such as when Odysseus attempts to hide his tears and succeeds at keeping them from everyone but Alcinous, who grasps (enoēsen) them (*Od*.8.533).[14]

Conversely, there are also numerous occasions in which noein is used to indicate deception, ignorance, or a failure to grasp the significance of some situation. If, as the scholarly tradition has assumed, noos and noein have the same basic semantic range, in these instances we should expect to find cases where someone happens to noein something in some way incorrectly, since we find cases where people are described as having a deceived, incomplete, or foolish noos. But neither Homer nor Hesiod uses noein in this fashion. With one possible exception (discussed below), they both uniformly say in these situations that someone did not noein: someone did not grasp the situation and did not get it correct. A representative example of such a use occurs at *Iliad* 19.112–13. Here, Hera tricks Zeus into a foolish oath:

ὣς ἔφατο· Ζεὺς δ' οὔ τι δολοφροσύνην ἐνόησεν,
ἀλλ' ὄμοσεν μέγαν ὅρκον, ἔπειτα δὲ πολλὸν ἀάσθη.

[Thus she spoke; and Zeus did not grasp (ou . . . enoēsen)
   her calculated guile,
but swore a great oath, and therein he was greatly deluded.]

And again at *Odyssey* 22.32 after the suitors say to the disguised Odysseus that he will be punished for accidentally, in their mistaken view, killing one of them, we are told that "they did not grasp (ouk enoēsan) in their childishness" (τὸ δὲ νήπιοι οὐκ ἐνόησαν) that doom was approaching them. Indeed, people who fail with respect to noein are often described as foolish or childish, nēpios. Even Achilles receives this designation at

Iliad 20.264. Aeneas drives his spear into Achilles's god-given shield, and Achilles grows alarmed that the shield will be pierced: "Childish man; he did not grasp (oud' enoēse) in heart or in spirit" (νήπιος, οὐδ' ἐνόησε κατὰ φρένα καὶ κατὰ θυμὸν) that his gift will not break.[15]

I could only find one possible exception to this overwhelming pattern within these texts. It occurs at *Iliad* 6.469–470, and the usage of noein found here is doubly atypical. Hector has returned from the day's warfare and meets with his wife and son. After speaking to Andromache, Hector reaches for Astyanax, who recoils from his father in fear,

ταρβήσας χαλκόν τε ἰδὲ λόφον ἱππιοχαίτην,
δεινὸν ἀπ' ἀκροτάτης κόρυθος νεύοντα νοήσας.

[. . . struck with terror seeing the bronze helmet and crest
    of horsehair,
grasping (noēsas) it nodding dreadfully from the topmost of
    the helmet.]

This passage is odd for multiple reasons. First, while what Hector's son is described as grasping here is not incorrect—the crest of horsehair is indeed nodding up and down—it nevertheless seems false to suggest that Astyanax has a good grasp of the situation in this scene. Indeed, what seems to be occurring in this passage is that he precisely does not grasp or recognize his father, and so gets scared. Throughout Homer and Hesiod, a scene like this would normally be expressed by saying that Astyanax does not noein his father, being a child. Indeed, that is the second thing that makes this usage so atypical: as discussed above, the word most commonly used to describe the person who does not noein something important is nēpios, childish.[16] There is no antecedent in the text for a child being able to noein anything at all.

As such, this passage is clearly a nonstandard outlier in Homer and Hesiod. It is not unambiguously nonveridical: perhaps the sense of this passage should be read as suggesting that the best grasp of a situation that can be expected of a child as young as Astyanax is not very penetrating. In any case, it is an abnormal and unprecedented usage for reasons besides the veridicality of the verb, and so there is every reason to think that it does not disclose the verb's standard sense. Given that this passage is such an anomaly—it is contradicted by every other analogous passage in Homer and Hesiod—by itself it does not suggest that the basic sense of noein is nonveridical.

Let us turn to secondary uses of the verb. As is the case with noos, noein is often used in contexts that point toward planning or deciding. As I suggested earlier, I see this usage as fundamentally continuous with the primary use, and the basic sense of the verb is usually something like "someone grasps what to do in this situation." It is almost always the case that this grasp of what to do leads directly to action, though there are exceptions.[17] In this usage, the verb is not indicating that someone is making a prediction, like Theoclymenus did when prophesizing the doom of the suitors. Instead, when someone grasps what they should do, their grasp actively determines or shapes what the situation will become in the future. For example, after telling the story of Meleager refusing to defend his city on account of a slight until the last moment and thus not receiving praise and gifts, Phoinix tells Achilles, "Don't grasp (noei) things that way with your phrēn," (ἀλλὰ σὺ μή μοι ταῦτα νόει φρεσί, *Il.*9.600). Phoinix does not want Achilles to decide the way that Meleager did, because if he does, things will indeed turn out for him how they turned out for Meleager. Conversely, if Achilles doesn't decide that way and decides to do something different, things will turn out differently. Something similar is true in the stock phrase used throughout the *Odyssey*:

ἔνθ' αὖτ' ἄλλ' ἐνόησε θεά, γλαυκῶπις Ἀθήνη.[18]

[Then the goddess, gleaming-eyed Athena, grasped something else to do (enoēse, planned or decided something else).]

In grasping what to do in the situation to realize her designs, Athena actively intervenes and thus changes the situation that she is grasping: for instance, she makes it so that the suitors are not able to murder Telemachus as they were getting ready to attempt.

Nevertheless, the veridical force of the verb's primary sense is preserved in this secondary usage. The plan or decision that someone grasps is always predicated upon a correct grasp of the current situation. In other words, it is always a sensible plan, given the situation, the character of the individuals formulating the plan, and their goals and purposes. This is true even when what someone is trying to accomplish is something wicked; for example, where Hesiod describes kings who pervert justice in their cities as "those who grasp how to do baneful things" (οἳ λυγρὰ νοεῦντες, *WD*.261). These kings do indeed succeed at perverting justice within their cities—they properly grasp the conditions within their cities and how to exploit those conditions for their own benefit—even though their actions are ultimately foolish and fail to heed the anger of the gods (*WD*.251). Conversely, at *Odyssey* 18.230 Telemachus says, "But I am not able to grasp a plan of action (noēsai) concerning all this with wisdom" (ἀλλά τοι οὐ δύναμαι πεπνυμένα πάντα νοῆσαι) because he cannot figure out how to deal with the suitors given his individual powerlessness.[19]

Finally, noein is sometimes used in contexts in which it seems to point to a general tendency to grasp situations or what to do, rather than to specific instances of such a grasp; for example, at *Iliad* 10.247 Diomedes says of Odysseus that περίοιδε νοῆσαι, which means something like that he knows well how to grasp things or exceeds others in his capacity to grasp things. These uses of the verb still preserve the veridical sense of its primary use, as in each case the tendency being described is the tendency to grasp them correctly.[20]

To summarize the results of this study of the uses of noein in Homer and Hesiod, the primary sense of the verb is to grasp a situation correctly. The verb also gets used in two secondary senses, a tendency to grasp a situation correctly and to grasp what to do given some situation. Both secondary senses preserve the veridical sense of the primary usage. Nearly anything can be the object grasped: (1) immediately present individuals and situations (thus tying the verb to what we moderns would consider to be perception), (2) the past and future, (3) the possibility of something, (4) significant absences, (5) conceptual distinctions, (6) a truth that is either obscured, concealed, or unclear (tying the verb to cognition), and (7) what to do given a particular situation (tying the verb to practical reasoning, volition, and action). When someone's individual grasp of the situation is incorrect in some regard, they are said to fail to noein, rather than to noein incorrectly, which again suggests that the verb's basic sense is veridical. Lastly, noein is not merely receptive or passive, but actively shapes the situation that it grasps by shaping the actions of the individuals who grasped it, who in turn shape the situation.[21]

Noos and Noein in Parmenides's Poem

Having surveyed the basic sense of noos and noein found in Homer and Hesiod, I will now argue that Parmenides uses noos and noein in a consistent manner, perhaps even widening the slight semantic gap between the noun and verb. Noos is used in contexts in which we would expect what is emphasized to be the particularity or partiality of some individual's grasp of the situation, and so is used in nonveridical contexts: indeed, in the extant fragments of the poem, it is mainly used in nonveridical contexts. Noein is used in contexts in which what is being emphasized is the link between grasping something and truth. Lastly, noēma, the content of someone's noos or that which was grasped in the act of the verb, fluctuates between the two senses depending on whether it is describing the content of someone's particular grasp or that which was grasped by noein, and so can be veridical or non-veridical depending on whether it is paired to noos or noein. The largest difference between Parmenides's usages and those of Homer and Hesiod is that what is being grasped is usually not some particular concrete situation, at least in the Truth portion of the poem, but instead something like the "situation" of reality as a whole, certain conceptual distinctions, or the impossibility of something. There were infrequent uses of the latter two in Homer and Hesiod, and the range of objects that can be grasped by noein is so large that using the verb to mark a grasp of "reality" is a consistent extension of its use.

Consider the use of noein in Fragment 2 (2.1–2):

Εἰ δ' ἄγ' ἐγὼν ἐρέω, κόμισαι δὲ σὺ μῦθον ἀκούσας,
αἵπερ ὁδοὶ μοῦναι διζήσιός εἰσι νοῆσαι

[Come now, I will tell you—and do preserve my story, when you have heard it—
about those ways of inquiry which alone there are to grasp (noēsai).]²²

There are two ways to read this use of noein: it is only possible to grasp these two ways of inquiry (and so these are the only ways), or these ways of inquiry are the ways to noein. I find the former the more natural way of reading this passage based on Homer's and Hesiod's uses of the verb, but either way noein would be veridical.²³ On the first reading, the youth is being asked to grasp a conceptual distinction: the way "it is" is distinct from the way "it is not," and that the former is the way of truth, and that the latter is fruitless and impossible. As such, it closely resembles Hesiod's use of noein at *Works and Days* 11–13, discussed previously, in which two different kinds of strife are distinguished and their differences grasped. Those who adopt the second reading see noein as the goal of the two ways, a goal that only the first, "it is," ever attains. "It is not" is a sterile way that doesn't arrive at any grasp, correct or otherwise.[24]

A related grammatical ambiguity confronts the equation of noein and einai in Fragment 3:

... τὸ γὰρ αὐτὸ νοεῖν ἐστίν τε καὶ εἶναι.

[For the same thing is for noein as is for einai.]

Again, in this passage there are two ways of reading the relation between noein and einai that Parmenides is positing. First, Parmenides could be saying that the verbs noein and einai mean the same thing or pick out the same activity, or he could be proposing that the subject matter of both verbs is the same: "things that are" are also "things that are grasped." However this ambiguity is resolved, noein has to be read as having a veridical sense for the passage to be coherent. Conversely, when read with a veridical sense, both interpretations make relatively plausible claims. Starting with the second interpretation, if noein means to grasp a situation correctly, then what the situation *is* and how the situation *is grasped* will indeed overlap in every case. The first interpretation—that the activity of grasping is the same "activity" as being—is not as obvious a connection, but it too has antecedents in the earlier tradition. As I previously suggested, noein does not merely have a passive or receptive sense, but rather plays a role in actively constituting the situation: it is because Zeus grasps (noēse) the danger of Typhoeus that the monster does not become king of gods and mortals (Hes.*Th*.820).[25] Certainly, Parmenides would be expanding and generalizing the connection between the situation and grasping it beyond anything suggested in earlier authors, but such a usage of noein would not be unprecedented.

The two ways of inquiry (and the corresponding grammatical ambiguity) occur again at Fragment 6 (6.1–9), which also features an important nonveridical use of noos.

Χρὴ τὸ λέγειν τε νοεῖν τ' ἐὸν ἔμμεναι· ἔστι γὰρ εἶναι,
μηδὲν δ' οὐκ ἔστιν· τά σ' ἐγὼ φράζεσθαι ἄνωγα.
Πρώτης γάρ σ' ἀφ' ὁδοῦ ταύτης διζήσιος <εἴργω>,
αὐτὰρ ἔπειτ' ἀπὸ τῆς, ἣν δὴ βροτοὶ εἰδότες οὐδὲν
πλάττονται, δίκρανοι· ἀμηχανίη γὰρ ἐν αὐτῶν

στήθεσιν ἰθύνει πλακτὸν νόον· οἱ δὲ φοροῦνται
κωφοὶ ὁμῶς τυφλοί τε, τεθηπότες, ἄκριτα φῦλα,
οἷς τὸ πέλειν τε καὶ οὐκ εἶναι ταὐτὸν νενόμισται
κοὐ ταὐτόν, πάντων δὲ παλίντροπός ἐστι κέλευθος.

[It is necessary to assert and grasp (noein) that this is Being.
　For it is for being,
but nothing is not. These things I command you to heed.
From this way of inquiry <I keep you> first of all,
but secondly from that on which mortals with no
　understanding
stray two-headed, for perplexity in their own
breasts directs their noos astray and they are borne on
deaf and blind alike in bewilderment, people without judgment,
by whom this has been accepted as both being and not
　being the same
and not the same, and for all of whom their journey turns
　backwards again.]

The precise rendering of 6.1–2 is highly contested,[26] but the basic sense of noein seems straightforward enough: depending on which reading of Fragment 3 is preferred, either we must grasp that being (or maybe "things") are rather than are not, or we are being told that noein (and also legein, "to speak") is the same as being. Either way, we are told to avoid the way of inquiry "it is not." While what was emphasized in Fragment 2 was the distinction between the two ways of inquiry, emphasized here is the nature of the two ways: the necessity of the first and the impossibility of second. However, the goddess then introduces either a third "way" of inquiry or a pseudo-way in which the distinction between "it is" and "it is not" is not grasped or maintained. The noos of those who fail to grasp this distinction goes astray. Such perplexed, "two-headed" individuals wander confusedly between the two ways of inquiry without grasping the distinction between them (and the impossibility of the second). This passage thus offers sharp contrast between the use of noos and noein in the poem, and the contrast conforms to Homer and Hesiod.

Continuing with the same idea, in Fragment 7 (7.1–2), the goddess says:

Οὐ γὰρ μήποτε τοῦτο δαμῇ εἶναι μὴ ἐόντα·
ἀλλὰ σὺ τῆσδ' ἀφ' ὁδοῦ διζήσιος εἶργε νόημα·

[For this principle shall never be vanquished, so as to allow
 things to be that are not,
but do keep the content of your grasp of things (noēma)
 from this way of inquiry.]

The noēma in question here is that of the youth the goddess is addressing. We can therefore read this passage as exhorting us to keep our noos free from the perplexity that leads humans astray into the pseudo-way that blends "it is" and "it is not" together, as discussed in Fragment 6 and as is implied by the rest of this fragment. As such, Parmenides's usage, which suggests the possibility of the noēma (the content of noos) being in error, and the goddess warning us to avoid that error, remains consistent with Homer and Hesiod.[27]

Along similar lines, the goddess says in Fragment 8 (8.7-9):

οὔτ' ἐκ μὴ ἐόντος ἐάσσω
φάσθαι σ' οὐδὲ νοεῖν· οὐ γὰρ φατὸν οὐδὲ νοητόν
ἔστιν ὅπως οὐκ ἔστι.

[I shall not allow you to say or to grasp (noein) that it grew
 'from not-being,' for it cannot be said (literally: "it cannot
 be the content of a saying") or the content of a correct
 grasp (noēton)
that anything is not.]

It is not permitted (i.e., it is impossible) to grasp that being grew from not-being, because not-being cannot be the content of an actual grasp. The noēton in this instance is clearly related back to the noein on the same line, and thus preserves the veridical force of the verb: the very sense that was lacking from noēma in Fragment 7 on account of it being linked to the youth's noos.

Noēma is used as the object grasped in noein again at lines 8.16–18:

ἔστιν ἢ οὐκ ἔστιν· κέκριται δ' οὖν, ὥσπερ ἀνάγκη,
τὴν μὲν ἐᾶν ἀνόητον ἀνώνυμον (οὐ γὰρ ἀληθής
ἔστιν ὁδός), τὴν δ' ὥστε πέλειν καὶ ἐτήτυμον εἶναι.

[Is, or is not? Now that it has been decided, as was necessary,
 to leave one way the content

of no correct grasp (anoēton) and nameless, since it not a real way, and for the other to be a way and authentic.]

The "it is not" way of inquiry produces no content that can actually be grasped or named, and so is anoēton: without possible, actually graspable noēma. Something analogous is at lines 8.34–36:

Ταὐτὸν δ' ἐστὶ νοεῖν τε καὶ οὕνεκεν ἔστι νόημα.
Οὐ γὰρ ἄνευ τοῦ ἐόντος, ἐν ᾧ πεφατισμένον ἐστιν,
εὑρήσεις τὸ νοεῖν·

[It is the same thing to grasp (noein) as is cause of the
  things grasped (noēma),
for you cannot find grasping (noein) without something that
  is, to which it is betrothed.]

Again, the passage has grammatical ambiguities, and the precise interpretation of Fragment 3 will shape that of 8.34. "The cause of the thing grasped" is usually taken, going back to Fragment 3, as being, which is supported by 8.35–36. Thus, whether the passage is about the subjects of noein and einai or the verbs themselves, the sense of noein is veridical. Indeed, this passage could be Parmenides directly stating the central claim of this chapter: that noein is unavoidably veridical, "betrothed"[28] to the actual situation that is the source of the noēma, what is correctly grasped.[29]

Finally, Fragment 16 is the one extant part of Doxa where either noos, noein, or noēma appears. Noein does not appear, but both nouns do.

Ὡς γὰρ ἕκαστος ἔχει κρᾶσιν μελέων πολυπλάγκτων,
τὼς νόος ἀνθρώποισι παρίσταται· τὸ γὰρ αὐτό
ἔστιν ὅπερ φρονέει μελέων φύσις ἀνθρώποισιν
καὶ πᾶσιν καὶ παντί· τὸ γὰρ πλέον ἐστὶ νόημα.

[For as is the temper which it has of the vagrant body at
  each moment,
so is noos present to men; for it is the same
awareness belonging to the nature of the body for all and
  each; for the preponderant is the noēma.]

The basic thrust of Parmenides's theory of human cognition in this passage has been articulated in the literature: human noos is the result of the mixing of two elements, usually thought to be the hot and the cold, and the noēma, the content of that noos, is determined by whichever of these elements is preponderant; for example, a "hot" noos is a grasp of hot things, and a "cold" noos is a grasp cold things. For our purposes, what is relevant is how this again emphasizes the particularity and partiality of an individual's noos. According to this theory, an individual's noos at the very least will tend toward being incomplete, as it will tend only to be a grasp of either the hot things or the cold ones, rather than a complete grasp of a situation (presumably, both hot and cold things).[30]

A survey of Homer, Hesiod, and Parmenides shows that noein has veridical force that noos does not necessarily possess. Parmenides's uses of the noun and verb are consistent with the earlier authors, which helps explain uses that otherwise seem at variance. I hope this chapter therefore supplies resources in the service of a full interpretation of the poem and the relation between noein and einai, to which Michael Wiitala turns our attention in the following chapter.

Notes

1. Von Fritz 1945, 236.
2. Von Fritz 1945, 239.
3. Studies of noos and noein consulted for this chapter include Heidegger 1935, chap. 4; von Fritz 1943 and 1945; Warden 1971; Sullivan 1980, 1988, 1990, and 1996; Lesher 1981 and 2008; Nagy 1983; and Benzi 2016.
4. Von Fritz 1943, 83. The way he characterizes this usage is as picking out the function in general, as opposed to the function at a particular time. I would add that I did not find any instances in Homer or Hesiod where someone is said to noein with, through, or in their noos—people are instead said to noein with their thumos or their phrēn.
5. Benzi 2016, 9 argues that this intuitive notion of noos is no longer operative in Parmenides and has been replaced by a partially deductive one. This conclusion seems to overstate the results of his study, however. His reading of Parmenides is such that the divergent mortal noos in Parmenides can be corrected by argumentation, unlike in earlier authors. Even if this claim is accepted, however, that only entails that noos (and noein) can be tested by argumentation, not that noos itself is fundamentally deductive.
6. The Greek text of *Works and Days* is taken from Evelyn-White 1914. Unless otherwise specified, translations are loosely based on those found in the same volume, heavily modified by me. Von Fritz (1945, 10) sees an ignorant noos as a unique innovation in Hesiod, though there seem to be cases in Homer

similar enough that I do not see the same discontinuity between the authors. See also Sullivan 1996, 31–51.

7. The Greek text of *Iliad* is taken from Monro and Allen 1920. Unless otherwise specified, translations are based on Alexander 2015, modified by me.

8. The Greek text of *Odyssey* is taken from Murray 1919. Unless otherwise specified, translations are loosely based on that same volume, heavily modified by me.

9. Though noos could also refer to Theoclymenus's grasp of this situation, which is also well formed compared to the suitors.

10. See also: Hom.*Il*.1.522, 2.391, 3.21, 3.30, 3.374, 4.200, 5.95, 5.312, 5.475, 5.669, 5.680, 5.711, 6.484, 7.18, 8.10, 8.91, 8.132, 11.248, 11.284, 11.343, 11.521, 11.575, 11.581, 11.599, 12.143, 12.331, 12.393, 15.395, 15.422, 15.453, 15.649, 17.116, 17.483, 17.486, 17.682, 20.291, 20.419, 21.49, 21.418, 21.527, 21.550, 21.563, 22.463; Hom.*Od*. 1.257, 4.116, 4.148, 4.653, 6.163, 7.290, 8.94, 8.271, 10.375, 12.245, 13.318, 15.59, 15.348, 16.283, 17.278, 19.232, 19.552, 20.204, 22.163, 24.232; Hes. *Sh*.410; Hes.*Th*.838.

11. See also Hom.*Il*.5.475, 9.104–105, 10.550, 15.81, 20.310, 23.415; Hom. *Od*.1.58, 5.188, 16.5, 18.228, 20.367, 21.257, 24.61.

12. See also Hom.*Od*.1.32, 16.160–64.

13. See also Hom.*Od*.15.170. On the relation between noein and signs, see Nagy 1983, 35–55.

14. See also Hom.*Od*.6.67, 17.301.

15. See also Hom.*Il*.5.665, 9.537, 10.501, 16.789, 19.112, 20.264, 22.445, 24.337; Hom.*Od*.7.37, 9.442, 11.62, 13.270, *Od*.22.32; Hes.*Th*.488; Hes.*WD*.89, 286, 484.

16. E.g., Hom.*Il*.5.20.264; Hom.*Od*.9.442, 18.228, 20.309, 22.32; Hes.*WD*.286, 484.

17. For example, Hom.*Il*.24.560, which I read as Achilles telling Priam that he grasps the situation in such a way that he intends to return Hector's body because he knows that it is the will of Zeus, but that he still might not do it if Priam doesn't stop pestering him.

18. Repeated at Hom.*Od*.2.382, 2.393, 4.795, 6.113, 18.187, and variants of the same construction are employed at 5.365, 23.242, 23.344.

19. See also Hom.*Il*.1.543, 1.549, 22.136, which I read as referring to the decision that Hector makes at the end of the prior monologue to stand and face Achilles, 22.235, 23.140, 23.193; Hom.*Od*.2.122, 3.26, 4.219, 6.251, 7.299, 16.409, 17.576. There are also a series of closely related uses in the *Odyssey* in which someone complements a piece of advice given by another by saying that the advice urges them to act with noein, with a grasp of what is good to do in the situation. These include *Od*.16.136, 17.193, and 17.281.

20. See also Hom.*Il*.1.577, 7.358 and 12.232, both of which I read as having the basic sense of "you generally know how to grasp what to say better than this,"

23.305; Hom.*Od*.5.170, which I read as Calypso claiming that the Olympian gods are better than her with respect to grasping what to do in situations and realizing their designs; Hes.*WD*.293-297.

21. Nagy 1983, 43-44 also notes that noos and noein have an active sense in Homer, where various signs are said to be decoded but also, in a few instances, encoded by the verb noein.

22. The Greek text of Parmenides's poem is taken from Diels. Unless otherwise specified, translations are based on those in Coxon 2009, modified by me.

23. Coxon's translation of Fragment 2 (2009, 290) suggests the first interpretation. For a historical overview of this and related grammatical disputes in Truth, see Giancola 2001.

24. For defenses of this second reading, see Robbiano 2006, 81-82 and Mourelatos 2008, 55-56 n. 26 and 75-78.

25. The Greek text of the *Theogony* is taken from Evelyn-White 1914. Unless otherwise specified, translations are loosely based on those of the same volume, heavily modified by me.

26. For an overview of all grammatical ambiguities confronting these lines, see Barrett 2004, 268-69.

27. Alternatively, Coxon (2009, 309-10) reads this passage as exhorting us to avoid the way "it is not." It seems strange to me to say that way produces noēma—it is said to be anoēton at 8.16-18—whereas the wayward noos seems to be described as having noēma at Fragment 16. Regardless, if Coxon is correct, noēma is still connected to the noos of the youth in this passage.

28. In this translation of πεφατισμένον as "betrothed," I deviate from the translation defended at Coxon 2009, 331-32. For a defense of this translation, see Mourelatos 2008, 170-72.

29. Because of space, I will not consider the use of noos at 4.1 or of noēma at 8.50. However, I see them as entirely consistent with my reading of both nouns.

30. For a good discussion of this passage, see Tor 2015.

Chapter 6

Noein and Einai in the Poem of Parmenides

MICHAEL WIITALA

The text of Fragment 3 of Parmenides's poem, . . . τὸ γὰρ αὐτὸ νοεῖν ἐστίν τε καὶ εἶναι,[1] can be construed in two basic ways: as asserting that (1) νοεῖν (to grasp, apprehend) and εἶναι (to be) are the same,[2] or that (2) the same thing is there for νοεῖν (grasping, apprehending) and for εἶναι (being).[3] Faced with this ambiguity and the parallel ambiguities in Fragments 2.2, 6.1–2, and 8.34–36, commentators have attempted to construe the text in either one way or the other, thereby generating significantly different interpretations of the Truth section as a whole.[4] In this chapter, I take a different approach. Instead of seeing the ambiguity as a liability,[5] I approach it as a resource, by offering an interpretation that combines both construals of the text. I argue that the same thing is there for νοεῖν and εἶναι precisely because νοεῖν and εἶναι are the same thing.[6] Νοεῖν, I argue, should be understood in Parmenides, as in Homer, as to grasp, to apprehend, or to be coherently oriented to a situation. Yet in Parmenides, unlike in Homer, the situation in question is that of grasping being (ἐόν).[7] Thus, Parmenides has the goddess affirm that the same thing—being—is there for both νοεῖν and εἶναι. Grasping being, however, is what being does in relation to itself: "being holds to being (τὸ ἐὸν τοῦ ἐόντος ἔχεσθαι)" (4.2; see also 8.25). The "to be" (εἶναι) that holds being to being is not a bond of bronze, rope, or some other material as in Homer (e.g., *Il.* 5.386–91; *Od.* 12.160–64, 194–200), but instead the intelligible grasp, apprehension, or orientation that is nothing other than νοεῖν. In the first section of this chapter, I explain and defend the interpretive hypothesis I will use to approach the ambiguity of Fragment 3. Then, in the next section, I develop my interpretation of νοεῖν. Finally, I defend my reading of εἶναι in light of the relationship between νοεῖν and the two ways of seeking that the goddess first identifies in Fragment 2.2.

Ambiguity in the Truth Section

While most commentators grant that there is intentional ambiguity in the Proem and/or Doxa of Parmenides's poem,[8] few acknowledge such ambiguity in the Truth section.[9] Instead, the typical approach to interpreting a passage in Truth is to disambiguate it. Yet as James Barrett has noted, Truth is rife with ambiguity. Moreover, he has convincingly argued that the ambiguity in Truth is intentional and has an educative function. The sort of educative function Barrett highlights is first and foremost that of struggle. By struggling with ambiguity, such as the polysemy of words like νοεῖν and εἶναι, we readers who engage in philosophy are forced to challenge customary ways of thinking and, in the words of the goddess, "judge by discourse the much-contested refutation uttered by me" (κρῖναι δὲ λόγῳ πολύδηριν ἔλεγχον ἐξ ἐμέθεν ῥηθέντα, 7.5–6). I want to suggest, however, that this is not the only sort of intentional ambiguity at play. There is also the kind of intentional ambiguity in which different construals of a passage operate together to express one underlying idea. Following Charles Kahn, I will call this latter sort of ambiguity "linguistic density."[10] Linguistic density, as Kahn uses the term, is "the phenomenon by which a multiplicity of ideas are expressed in a single word or phrase"[11] or "the use of lexical and syntactic indeterminacy as a device for saying several things at once."[12] A linguistic density is such that all construals taken together provide a more complete insight than each does individually.

Poets throughout history have employed linguistic density, and some commentators have acknowledged its presence in Parmenides.[13] Barrett, for example, proposes that the syntactic ambiguity in Line 1 of the Proem, concerning whether the relative clause modifies what precedes it or follows it, is the sort of ambiguity I am calling "linguistic density." The text reads:

ἵπποι ταί με φέρουσιν ὅσον τ' ἐπὶ θυμὸς ἱκάνοι
πέμπον, ἐπεί μ' ἐς ὁδὸν βῆσαν πολύφημον ἄγουσαι
δαίμονος . . .

[The mares that carry me as far as the heart might desire were
escorting me, when they had brought and put me on the path rich in discourse
of the goddess . . .][14]

As Barrett points out, "The ambiguity [of the relative clause in line 1.1] is instructive in that it points to ways in which both (types of) journeys—past and present—are identical: guided by the same horses, both alike conform to the reach of the thumos."[15] My contention is that the syntactic ambiguity of Fragment 3 is also an instance of linguistic density, as are the related ambiguities in Fragments 2.2, 6.1–2, and 8.34–36, although I will only discuss 2.2 and 3 in this chapter.

Many commentators, however, hold that while fragments like 2.2, 3, 6.1–2, and 8.34–36 seem ambiguous to us, their syntax was in fact not ambiguous to Parmenides's original audience. There is no syntactic ambiguity in these passages, so the story goes. Instead, on this interpretive hypothesis, the ambiguity we see is due to the fact that we are importing syntactic possibilities from later and/or earlier Greek into the text as we read it, possibilities that simply did not exist in the Greek of Parmenides's original audience.[16] Yet the fact that there is still widespread scholarly disagreement among those who approach the poem with this interpretive hypothesis indicates that extant textual evidence is insufficient to support it or warrant its continued use.[17] Moreover, the hypothesis is rather dubious from the start given the poetic medium,[18] the journey and initiation ritual motifs,[19] and the goddess's instructions to the kouros, which also seem to be directed to the reader.[20] She tells the kouros to restrain his grasp of things (νόημα, 7.2), not to let the force of custom carry him (7.3), and so on. Carefully constructed textual ambiguities in a poem are an effective way of drawing readers away from customary habits of thinking and speaking. The reason so many modern scholars have assumed the Truth was unambiguous to its original audience seems more due to post-Platonic assumptions about what philosophically rigorous writing should be than to anything in the text of Parmenides or other early Greek philosophers.[21]

Consequently, I think my alternative interpretive hypothesis—that the syntactic ambiguity in Fragment 3 and elsewhere in Truth is intentional and linguistically dense—is safer and has more textual warrant than the hypothesis that Truth was unambiguous to its original audience.

Noein

Recall the text of Fragment 3: . . . τὸ γὰρ αὐτὸ νοεῖν ἐστίν τε καὶ εἶναι. The syntactic ambiguity concerns whether τὸ αὐτό is the subject or predicate of ἐστίν. If τὸ αὐτό is taken as the predicate, the νοεῖν and εἶναι used substantively are the subject: ". . . for it is the same to grasp and to be" or ". . . for to grasp and to be are the same."[22] If, in contrast, τὸ αὐτό is taken as the subject, we presumably have a dative use of νοεῖν and εἶναι that enables τὸ αὐτό to serve simultaneously as the object of νοεῖν and subject of εἶναι: ". . . for the same thing is there to grasp and to be" or ". . . for the same thing is for grasping and for being."[23] Instead of attempting to eliminate this ambiguity, my approach will be to let it guide us toward an understanding of the νοεῖν and εἶναι the goddess speaks of in Truth. Our goal will be to find a way of understanding νοεῖν and εἶναι compatible with and revealed by both construals. Thus, we will need to understand νοεῖν and εἶναι such that νοεῖν and εἶναι are the same thing, and such that the same thing is there for νοεῖν and εἶναι. My contention is that both νοεῖν and εἶναι name the "hold" (ἔχεσθαι, συνέχεσθαι) that the goddess claims being has in relation to being (τὸ ἐὸν τοῦ ἐόντος; see esp. 4.2, 8.23). I argue that νοεῖν and εἶναι are the same because they are the same hold of being to being; whereas the same thing is there for νοεῖν and εἶναι in that being is there for νοεῖν and εἶναι to hold.

On the side of νοεῖν, this hold is the way the various entities in a given context veridically fit together. In Homer, as Kurt von Fritz's well-known study has shown, the basic sense of the word νοεῖν is "to realize a situation."[24] "The situation," according to von Fritz, "is the real object of the mental act designated by the verb νοεῖν."[25] To realize a situation is to grasp how the various entities and factors relevant to that situation really hold together. In Homer, the situation is generally a practical one and the relevant entities are gods, people, and other factors pertinent to making a sagacious decision. Νοεῖν is the realization that supervenes upon one's perception of these entities and other situational factors so that one grasps how they really fit together and what is in fact practically relevant.

Although I think von Fritz's characterization of νοεῖν in Homer as "to realize a situation" is right, his categorization of νοεῖν as a "mental act" is misleading. To realize or to grasp a situation in the ways described by νοεῖν in Homer is something one undergoes with the whole of one's being: cognitively, perceptually, emotionally, practically, physiologically, and ontologically. It requires taking one's place within and adopting a practical stance toward a situation. Or, in the case of situations in which one is not presently enmeshed, νοεῖν requires taking what would be one's place(s) within and adopting what would be one's stance(s) toward those situations. This explains why νοεῖν also has the sense of "to plan." To grasp a possible future situation is to be coherently oriented to it in the sense of having a sagacious plan; whereas to grasp a situation in the past or present is to be coherently oriented to it in the sense of realizing its character and how to respond to it. Put differently, to realize a situation is for one's whole being to be coherently oriented to the beings whose holding together constitutes the situation as a situation.

Since a situation is only a situation insofar as the beings within it are coherently oriented toward one another, the result is that νοεῖν is the "hold" that constitutes a situation as a situation. I am not claiming that situations are solely constituted by mental acts. Instead, I mean to say that a situation is constituted by the coherent ways of being oriented, relative to one another, of the beings that compose the situation. Further, the coherent orientation of these beings toward one another is what constitutes the situation as a situation. If there are no beings oriented to one another in a coherent way, there is no situation in the relevant sense, since by a "situation" here we do not mean just any state of affairs, but instead a normatively charged one.

Far from being merely a mental act, νοεῖν in Homer and other early Greek writers is an act of one's whole being by which one takes one's place in and becomes co-constitutive of the situation to which one is coherently oriented. When Menelaus and Paris become noetically aware of one another in battle near the beginning of *Iliad* Book 3, first Menelaus, and then Paris, become coherently oriented to the situation (*Il*.3.21: τὸν δ' ὡς οὖν ἐνόησεν ἀρηΐφιλος Μενέλαος and *Il*.3.30: τὸν δ' ὡς οὖν ἐνόησεν Ἀλέξανδρος θεοειδής). As each in turn becomes coherently oriented—as each ἐνόησεν, realizes, or is noetically aware of the other—the situation itself is constituted by and coheres in their respective orientations or realizations.[26] The situation described in *Iliad* 3.15–37, and that coheres as Paris falls back and hides among his men in fear, is co-constituted by the noetic awareness that each has of the other's presence. The νοεῖν that Menelaus and Paris undergo is not simply a mental act, nor is it a response to a situation that exists independently of the response. Rather the νοεῖν they undergo is itself constitutive of the situation. If Menelaus and Paris had not become coherently oriented to one another, then the situation would not have been the situation that it was. Instead, it would have been a different situation.

As Paul DiRado argues in his chapter in this volume, correctly grasping a situation is the sense of νοεῖν in Homer, Hesiod, and Parmenides. Hence DiRado's analysis differs, I think rightly, from that of von Fritz. According to von Fritz, νοεῖν in Parmenides begins to have the sense of "logical reasoning" or "discursive thinking" not found in earlier authors. Von Fritz claims this warrants its translation in Parmenides as "to think," although he is careful to note that the sense of νοεῖν as intuitive is still primary.[27] The main reasons von Fritz cites for holding that νοεῖν has the sense of logical reasoning in Parmenides are (1) that νόος and νόημα can err in Parmenides,[28] and (2) that inference markers, such as γάρ (for), ἐπεί (because, when), οὖν (therefore), τοῦδ' εἵνεκα (on account of this), or οὕνεκα (wherefore) occur "in almost every sentence."[29] As DiRado shows, however, von Fritz mistakenly assumes that the words νόος, νόημα, and νοεῖν have the same semantic range. In Parmenides, as in Homer and Hesiod,[30] νόος and νόημα can err in the sense of partially grasping the truth.[31] Νοεῖν, by contrast, is always veridical.[32] This difference in the semantic range of νόος and νόημα, on the one hand, and νοεῖν, on the other, parallels the way it is natural to say in English that someone has a mistaken understanding of something but unnatural to say that someone understands something when she is in fact mistaken. If we allow for the meanings of νόος, νόημα, and νοεῖν to diverge in this way, then in Parmenides νοεῖν can simply mean what it meant in Homer and Hesiod: to grasp or be coherently oriented to a situation.[33] The difference will be that in Parmenides the relevant situation is every situation, which is to say being or intelligibility as such, or any situation or context that is there for speaking (λέγειν). Put differently, the relevant situation is the only situation there is: the situation of beings coherently oriented toward one another.

Einai

With this initial sketch of the meaning of νοεῖν in place, we can now move to εἶναι, which will in turn enable us to delimit further what the goddess means by νοεῖν. My contention is that both νοεῖν and εἶναι are the hold (ἔχεσθαι, συνέχεσθαι) that being has in relation to being. This hold is first described as such in Fragment 4:

> λεῦσσε δ' ὅμως ἀπεόντα νόωι παρεόντα βεβαίως·
> οὐ γὰρ ἀποτμήξει τὸ ἐὸν τοῦ ἐόντος ἔχεσθαι
> οὔτε σκιδνάμενον πάντηι πάντως κατὰ κόσμον
> οὔτε συνιστάμενον.

> [Look at things that being far off are nevertheless by understanding firmly present.
> For you will not cut off being from holding to being,
> Neither altogether scattered in every way in regular order
> Nor gathered together.] (4.1–4)

For sense perception, things are more or less temporally and spatially distant from the perceiver. Likewise, things are gathered or scattered relative to some spatial or temporal location. The goddess, however, asks the kouros to look at what is made firmly present by means of understanding (νόωι).[34] Indeed, she warns him in Fragment 7 against an aimless eye, ringing ear, and tongue. The goddess in Fragment 4 is pointing out to the kouros that things absent to the senses can be made firmly present by understanding. She wants him to notice a kind of hold that does not depend on spatial or temporal proximity: the hold of being to being.

What sort of hold is this that requires no spatial or temporal proximity? Fragment 8 offers guidance for answering this question. We can begin by considering 8.12–16:

> οὐδὲ ποτ' ἐκ μὴ ἐόντος ἐφήσει πίστιος ἰσχύς
> γίγνεσθαί τι παρ' αὐτό· τοῦ εἵνεκεν οὔτε γενέσθαι
> οὔτ' ὄλλυσθαι ἀνῆκε Δίκη χαλάσασα πέδῃσιν,
> ἀλλ' ἔχει· ἡ δὲ κρίσις περὶ τούτων ἐν τῶιδ' ἔστιν·
> ἔστιν ἢ οὐκ ἔστιν . . .

> [Nor from not-being will the strength of trust ever impel
> Anything to come to be beside it [viz. being]. Therefore neither to come to be
> Nor to perish did Justice allow, loosening the fetters
> But she holds [it/them]. And the judgment about these things belongs to this:
> __is__ or __is not__. (8.12–16)]

I follow Alexander Mourelatos in reading the ἔστιν (is) and οὐκ ἔστιν (is not) of the ways of searching, introduced in Fragment 2, as the copula with subject and predicate suppressed.[35] Mitchell Miller has, I think, correctly identified the purpose of the suppression. Miller argues that the effect of the suppression of subject and predicate "is to reverse the usual order of the conspicuous and the inconspicuous . . . the goddess brings the normally inconspicuous 'is' to the front and centre and challenges us to reflect upon it."[36] Although individual beings are in the forefront of our everyday mortal attitude, the goddess, by contrast, highlights the connection between individual beings, which she characterizes as the fetters that hold being to being (esp. 8.12–32). In 8.12–13, however, what the fetters hold is ambiguous. The wording allows the fetters to hold (i) being (ἐόν from 8.3), (ii) not-being (μὴ ἐόν from 8.12), or (iii) coming to be and perishing (γενέσθαι and ὄλλυσθαι in 8.13–14). The interpretation of the "fetters" I defend will allow us to say that they hold all three.

The fetters, I argue, are the __is__ and __is not__ to which the judgment belongs, or the two ways of inquiry from Fragment 2. In Fragment 2, the goddess asks the kouros to try to speak (λέγειν) and to grasp (νοεῖν) __is__ and __is not__. I take it that Parmenides intends for his readers to try and do the same. Yet when we try to speak or grasp __is__ and __is not__, we come to recognize (cf. φράζεσθαι, 6.2) that in the case of __is not__ this cannot be accomplished (see esp. 2.7, 8.8–9). It is in this recognition that we grasp being. We come to realize that being is whatever is such that it can be grasped and "is" can be said of it,[37] by trying and failing to speak and grasp that which is other than being, namely, not-being.

What prevents us from speaking or grasping not-being? On the one hand, the goddess suggests that the fetters (πέδαι) or bonds (δεσμοί) held by Justice, Necessity, and Fate prevent us. On the other hand, she characterizes __is__ and __is not__ as what prevent us from speaking or grasping not-being. The __is__ will not let us speak or grasp not-being, because not-being is not something about which any statement can be made. Yet neither will the __is not__ let us speak or grasp not-being, since, given that no statement can be said of not-being, we cannot even say, "Not-being is not," without contradiction. Even in saying that "no statement can be made about complete not-being" we contradict ourselves, since we are treating complete not-being as something intelligible and about which we can speak. Being, therefore, is the only thing that can be grasped and said. The __is__ constrains us to grasp and say being, because it holds being to being. Likewise, the __is not__ constrains us to grasp and say being, due to its failure as a way of grasping and saying anything at all. Since the fetters or bonds, on the one hand, and the __is__ and __is not__, on the other, are what prevent us from speaking or grasping not-being, we have good reason to think that the fetters or bonds just are __is__ and __is not__.

If the fetters in Fragment 8 are __is__ and __is not__, then what the goddess means by claiming that Justice holds being, not-being, coming to be, and perishing begins to become clear. On the one hand, the __is__, which is one of the fetters, constrains being by requiring anything we can speak or grasp to be a being: something such that "is" can be said of it. On the other hand, the __is not__, the other fetter, constrains not-being, coming to be, and perishing, by preventing us from speaking or grasping them. This is the sense in which the judgment in 8.15–16 belongs to __is__ and __is not__. By traveling down __is__ and __is not__ as ways of seeking to grasp something, one is faced with a judgment to make as they diverge. In making that judgment in the way Necessity demands (ὥσπερ ἀνάγκη, 8.16), we come to see that __is__ and __is not__ are the fetters holding being to being and excluding not-being.

That the judgment belongs to __is__ and __is not__ is already implicit in the claim of Fragment 2 that __is__ and __is not__ are ways of seeking νοῆσαι (2.2). As I say above, I read νοεῖν (of which νοῆσαι is the aorist infinitive) here and throughout as "to grasp a situation" or "to be coherently oriented to a situation," where the "situation" in question is every situation, or put differently, the only situation that ever is: a situation in which intelligibility shows itself and being holds to being. Yet even if one were to grant this meaning of νοεῖν, the question of how to interpret it within the context of 2.2 remains, due to an ambiguity similar to that of Fragment 3. The text of 2.2 reads: αἵπερ ὁδοὶ μοῦναι διζήσιός εἰσι νοῆσαι. The ambiguity concerns whether the ways of seeking (ὁδοὶ διζήσιος) are, logically speaking, the object of νοῆσαι,[38] or whether the object is left unspecified.[39] In other words, the ways of seeking can be construed either as ways that are there to be grasped by someone or as ways that someone can grasp something. For this reason, commentators have framed the difference between these two construals in terms of whether νοῆσαι—which is active in form—has an active or passive meaning in 2.2.[40]

On the first construal, νοῆσαι has a passive meaning: the only ways to be grasped. On the second construal, it has an active meaning: the only ways to grasp something. The second construal, however, gives rise to a further ambiguity. The goal of the ways of seeking can be construed, on the one hand, as to grasp something, or, on the other hand, as the thing to be grasped. In the first case, the goal of the seeking and the object of the grasping are construed as different from one another: the goal of the seeking is *grasping x*, while the object of the grasping is *x*. In the second case, the goal of the seeking and the object of the grasping are the same thing, *x*, with the result that the seeking and grasping are identified. Altogether, then, we have three basic construals of 2.2: (1) The ways of seeking are the object of νοῆσαι: "Which ways of seeking alone there are to be grasped;"[41] (2) The logical object of νοῆσαι is unspecified and the goal of both ways of seeking is νοῆσαι: "Which ways of seeking are the only [ways of seeking] to grasp [something]" or "Which ways of seeking alone are there for the purpose of grasping [something];"[42] (3) The logical object of νοῆσαι is unspecified and the goal of the ways of seeking is the object of νοῆσαι: "Which ways of seeking are the only [ways] to grasp [something]" or "Which ways of seeking are the only [ways] of grasping [something]."[43]

My approach is to understand the passage such that all three construals are true and informative.[44] First, consider the compatibility of construals (2) and (3). To seek to grasp something is to seek the thing to be grasped. Likewise, given that the thing to be grasped can only be had by grasping it, to seek the thing to be grasped is to seek to grasp it. Put differently, if we understand νοῆσαι as "to be coherently oriented to a situation," then seeking to be coherently oriented toward a situation is seeking to orient oneself coherently to that situation. But seeking to orient oneself coherently to a situation is nothing other than seeking to be in that situation as a situation to which one is coherently oriented. Hence, construals (2) and (3) are fully compatible.[45]

Yet the difference between (2) and (3) reveals something important. The ways of seeking, as the goddess says, are __is__ and __is not__ (2.3, 2.5). On construal (3), __is__ and __is not__ are ways of grasping or orienting oneself to something. On construal (2), __is__ and __is not__ are ways of seeking one can employ to grasp or orient oneself to something. On construal (2), therefore, someone can seek to be coherently oriented to something either by way of __is__, in which case what is grasped will be what-is or being (ἐόν); or by way of __is not__, in which case what is grasped would be what-is-not or not-being (μὴ ἐόν), if what-is-not or not-being could be grasped. Yet of course what-is-not cannot be grasped. Thus, on construal (2), according to which to grasp something is the goal of seeking, __is not__ has the status of a way seeking. As a way of seeking, __is not__ is a seeking to grasp something that fails to attain its goal.⁴⁶ On construal (3), in contrast, according to which the seeking and grasping are identified, __is not__ fails to be a way of seeking because __is not__ does not grasp anything. Hence, the difference between construals (2) and (3) reveals the sense in which __is not__ is a way of seeking (sense [2]) that nevertheless fails to be a genuine (ἀληθής) way of seeking (in sense [3]; cf. 2.2 with 8.17–18).

What about construal (1), according to which the ways of seeking are the object of νοῆσαι? On construal (1), the goddess is asking the kouros to orient himself to the only two ways of seeking that are there to be grasped or toward which one could orient oneself. If we grant that construals (2) and (3) are both true—the goddess is introducing the only ways of seeking to grasp something—and if we grant that anything insofar as "is" can be said of it is a being and can be grasped, then the only ways of seeking to grasp something, since we are speaking of them, must themselves be beings that can be grasped.⁴⁷ Therefore, one dimension of the task the goddess appoints to the kouros is for his grasping to grasp itself or for him to become coherently oriented to the only ways to be coherently oriented. Part of becoming coherently oriented to these sole ways of being coherently oriented is to recognize that and how (the two senses of ὡς/ὅπως at 2.3, 2.5, 8.2, 8.3, 8.9)⁴⁸ __is__ is a way of seeking that can be accomplished (ἀνυστόν), while __is not__ is a way that cannot be accomplished (2.6–8). This will require the kouros to grasp that and how __is__ is the only possible way to grasp something. The goddess introduces the two ways of seeking and asks the kouros to grasp them. If we preserve the ambiguity of construals (2) and (3), she asks the kouros to grasp the ways of seeking as the only two ways of seeking that are (ways of seeking / ways) to grasp something. Yet to grasp them as the only ways of seeking that are there to grasp something entails differentiating construals (2) and (3).

To grasp the ways of seeking entails understanding that __is__ is a way of seeking that is there to grasp something on both construals: it is both a way of seeking to grasp something (sense [2]) and a way to grasp something (sense [3]). Likewise, to grasp the ways of seeking entails understanding that __is not__ is a way of seeking that is there to grasp something only on construal (2), and not on construal (3), since __is not__ is not a way of grasping something, although it is a way of seeking to grasp something. Attaining these insights, in turn, should prompt the kouros to see how __is not__ "is not a genuine way" (οὐ γὰρ ἀληθής ἔστιν ὁδός, 8.17–18): it is not a way on construal (3); it is only a way on construal (2). This, in turn, should prompt him to conclude that __is__ is the only genuine way to grasp something. Put otherwise, the goddess is asking for the νοεῖν (to be coherently oriented, to grasp) of the kouros to be coherently oriented to or to grasp its own identity as the __is__ that holds being to being.

Hence, we arrive at the claim that νοεῖν and εἶναι are the same thing. The fetters or bonds by which Justice, Necessity, and Fate hold being to being are not made of bronze or some other material, but are rather the intelligible ways that beings are oriented toward one another. We can call those bonds ἔστιν and οὐκ ἔστιν, "__is__" and "__is not__." Furthermore, the bond we call ἔστιν or εἶναι can also be called νοεῖν, "to be coherently oriented." Every being, insofar as it is, is coherently oriented to every other. Both εἶναι and νοεῖν name the grasp or hold that being has in relation to being. Returning to Fragment 3, therefore, we can affirm that the same thing, namely being (ἐόν), is there for νοεῖν and εἶναι because νοεῖν and εἶναι are the same. It is the same to be coherently oriented and to be, because what it is for something to be is to be oriented in some coherent way to other beings.

Conclusion

I have argued that the syntactic ambiguities of Fragments 2.2 and 3 should be understood as intentional ambiguities and instances of linguistic density. The ambiguities not only prompt the reader to struggle with what the goddess means but also help guide the resolution of that struggle by their polysemy. If we approach such ambiguous passages with the interpretive hypothesis that the various ways they can be construed are intentional and reveal something the goddess wants to communicate, then we come to see the ambiguity as a resource rather than a hindrance for interpreting the poem and achieving the insights it offers. I hope I have shown that this is a fruitful way of approaching Fragments 2.2 and 3. Such an approach guides us to see the __is__ and __is not__ both as ways of grasping and as ways to be grasped. Further, it guides us to see them as the fetters or bonds that hold being to being. I think the interpretation I have offered here could be fruitfully extended to Fragments 6.1–2 and 8.34–36. The syntactic ambiguities in those passages are more extensive. Yet I expect that approaching them as instances of linguistic density would further reveal the insights into which Parmenides wishes to initiate his readers.

Notes

1. Unless otherwise indicated, the Greek text used is that of Diels and Kranz (1956). Translations are my own unless otherwise indicated.

2. Clement, Plotinus, and Proclus, our sources for Fragment 3, all construe the passage in this way. For modern commentators who prefer this construal, see e.g., Diels and Kranz 1956, 1:231; Verdenius 1942, 33–41; Kahn 1969, 721; Long 1996, 125–51; Sedley 1999, 120; Giancola 2001, 635–53; Henn 2003, 53–54; and Cordero 2004, 192.

3. For commentators who prefer this construal, see e.g., Zeller 1881, 584 n. 1; Burnet 1930, 173 n. 2; Tarán 1965, 41–44; Guthrie 1965, 14; Gallop 1984, 57; Coxon 2003, 210–12 and 2009, 296–97; and Graham 2010, 213, 236.

4. For an account of the interpretive history of Fragment 3, see Altman 2015, 197–230.

5. For the view, common among modern commentators, that the ambiguity in Fragment 3, and ambiguity in the Truth generally, is a liability, see e.g., Mourelatos 2008, xxi; and Altman 2015.

6. Perl (2008, 2014, 14–15) and Robbiano (2006, 58) read Fragment 3 in this way, but neither offers a detailed defense of this interpretation.

7. Cf. Long 1996.

8. By speaking of "intentional ambiguity" I do not mean to make a historical claim about Parmenides's authorial intentions. Instead, following Kahn (1979, 91), I use the expression simply to reflect that "we can construe an ambiguity in the text as meaningful only if we perceive it as a sign of the author's intention to communicate to us some complex thought."

9. On intentional ambiguity in Truth, see Barrett 2004 and Robbiano 2006, esp. 14–27.

10. Kahn 1979, 89–92. Kahn employs the term in the context of explaining the hermeneutic principles used in his interpretation of Heraclitus. Robbiano, adopting the term from Kahn, employs it in reading Parmenides (Robbiano 2006, 26–27). What Kahn calls "linguistic density" is second-type ambiguity in Empson 1966, chapter 2. Empson describes it as an ambiguity in word or syntax "when two or more meanings are resolved into one" (Empson 1966, 48).

11. Kahn 1979, 89.

12. Kahn 1979, 91.

13. Most of Empson's examples of what I am calling "linguistic density" (Empson's second-type ambiguity) are taken from British literature (Empson 1966, chap. 2). Stanford (1939, esp. 92) sees Empson's second-type ambiguity present

in ancient Greek literature, yet he does not consider ambiguity in Parmenides in any detail. Barrett (2004) and Miller (2006) have identified what I am calling "linguistic density" in the Proem of Parmenides's poem. Robbiano (2006, esp. 25–27, 31) sees intentional ambiguity and linguistic density throughout the poem, including in Truth. Mansfeld (1995, 228–32) argues that there is intentional ambiguity in the Proem and Barrett (2004) defends the view that there is intentional ambiguity in Truth as well.

14. The text and translation here are those of Barrett (2004, 270), who, to emphasize the ambiguity, omits the commas typically inserted into the Greek text to mark off the relative clause.

15. Barrett 2004, 270.

16. See Coxon 2003, 210–211 for a good example of this line of argument applied to Fragment 3.

17. The significant disagreement among scholars who adopt this interpretative assumption with respect to Fragment 3 is evident in the contrast between Giancola 2001 and Coxon 2003.

18. Robbiano 2006, chapters 1–2 contain insightful discussion of the poetic medium and genre.

19. Mary Cunningham discusses the initiation ritual motif in her chapter in this volume.

20. See Barrett 2014.

21. I say "post-Platonic" because Plato's dialogues give us no reason to think that philosophical writing should be free of ambiguity: quite the opposite.

22. For a detailed discussion of the syntax, see both Giancola 2001 and Coxon 2003.

23. Monro (1882, sec. 231) discusses this sort of "harsh change of subjects" in Homer, citing *Il.* 9.230 and *Od.* 2.226.

24. Von Fritz 1943, 79–93 and 1945, 223.

25. Von Fritz 1943, 85.

26. As von Fritz (1943, 85) points out, "It is nothing but an abbreviated and more concentrated expression when the description of the situation is replaced by the object in which the situation is focused."

27. Von Fritz 1945, 236–42, esp. 241–42.

28. Von Fritz 1945, 237–42.

29. Von Fritz 1945, 241.

30. See DiRado's chapter in this volume.

31. See frs. 6.5–6, 7.2, 8.50, 16.

32. See DiRado's chapter in this volume; cf. Robbiano 2006, 128–29.

33. For other accounts of the relationship between νοεῖν in Homer and Parmenides, see von Fritz 1945, 236–42; Guthrie 1965, 17–19; Mourelatos 2008, 68–70 and 164; Robbiano 2006, 50–59 and 128–32; Fronterotta 2007, 12–18; and Benzi 2016, 1–18.

34. I read νόωι here as a causal or instrumental dative, such that understanding (νόος) is what makes the absent things present. For a defense of this reading, see Coxon 2009, 306. For an alternative, see Fronterotta 2007, 13ff.

35. Mourelatos 2008, 55; cf. Owen 1960, 95; Guthrie 1965, 14; Cordero 2004, 51–53, 61–63; and Bredlow 2011, 283–84.

36. Miller 2006, 3.

37. Owen 1960, 95; cf. Guthrie 1965, 14–17.

38. For commentators who adopt this construal, see e.g., Burnet 1930, 173; Tarán 1965, 32; Guthrie 1965, 13–14; Kirk et al. 1983, 245; Coxon 2003, 211; and Graham 2010, 213 and 236.

39. For commentators who adopt this construal, see e.g., Mourelatos 2008, 55 n. 26; Kahn 1969, 703 n. 4; Giancola 2001, 642 and 649 n. 10; and Cordero 2004, 40–42.

40. See e.g., Giancola 2001, 640–49.

41. For a defense of (1), see Coxon 2009, 290. Syntactically, this construal seems to require the dative use of νοῆσαι (cf. Coxon 2003, 211).

42. For a defense of (2), see Robbiano 2006, 81–82; cf. Cordero 1984, 49–50. Syntactically, this construal requires νοῆσαι to be a final infinitive.

43. For a defense of (3), see Mourelatos 2008, 55 n. 26 and 75 n. 4; Kahn 1969, 703 n. 4; Giancola 2001, 642 and 649 n. 10; and Cordero 2004, 40–42. Various syntactic possibilities can support this. Cordero argues νοῆσαι as a final or consecutive infinitive supports this construal; but so does νοῆσαι read as an infinitive restricting the "sphere of action" of seeking to the realm of νοῆσαι. Likewise, the dative use of νοῆσαι as an instrumental dative could support (3).

44. Robbiano (2006, 81–82) takes this approach, but only identifies (1) and (2), not (3).

45. Mourelatos (2008) seems to grant this in reading νοῆσαι (2.2) as a final infinitive, suggesting (2), but claims that reading it as final "tends to identify thinking and the routes," which is (3).

46. Cf. Robbiano 2006, 82.

47. This is compatible with the goddess's claim at 8.17 that the way of __is not__ is ungraspable and unnamable (ἀνόητον ἀνώνυμον). The way of __is not__ is ungraspable and unnamable as a way of grasping, but it is graspable and namable as a way of seeking to grasp something.

48. Cf. Mourelatos 2008, 70–71; Robbiano 2006, 82–84.

Chapter 7

How Many Roads?

MATTHEW EVANS

In Parmenides's Fragment 2, the goddess[1] explicitly specifies two "roads of searching that are for thinking" (ὁδοὶ . . . διζήσιός εἰσι νοῆσαι): one that she associates with being (2.3), and another that she associates with not-being (2.5). But does she refer, anywhere else in the poem, to a *third* road: one that she associates with *both* being *and* not-being? My aim in this chapter will be to establish that, despite the various concerns raised in recent years by interpreters such as Néstor-Luis Cordero, Alexander Nehamas, Patricia Curd, and others,[2] the answer to this question is "yes": the goddess acknowledges three roads, not just two. As I hope to show, the consequences of this discovery are far from trivial; on the contrary, they place a number of powerful constraints not only on our reading of Fragment 2, but also on our understanding of the poem as a whole.

The Allure of the Two-Road Interpretation

It is tempting to imagine, after a quick look at Fragment 2 by itself, that the two-road interpretation must be correct. For consider what the goddess says there:

> But come now, I will tell you . . .
> the only roads of searching that are for thinking (ὁδοὶ
>   μοῦναι διζήσιός εἰσι νοῆσαι): [2]

the one, that it is and that it is not possible for it not to be
(ἡ μὲν ὅπως ἔστιν τε καὶ ὡς οὐκ ἔστι μὴ εἶναι), [3]
is a path of Persuasion, for she accompanies Truth; [4]
the other, that it is not and that it is right for it not to be
(ἡ δ' ὡς οὐκ ἔστιν τε καὶ ὡς χρεών ἐστι μὴ εἶναι), [5]
this indeed I point out to you to be an utterly uninformative trail; [6]
for neither could you ascertain what-is-not (τό γε μὴ ἐόν),
for that cannot be accomplished, [7]
nor could you point it out.[3] [8]

Here the goddess makes perfectly clear that, in her view, there are at least two roads of searching that "are for thinking" (εἰσι νοῆσαι): the one she specifies in line 3 (which I will call **the road of being**) and the one she specifies in line 5 (which I will call **the road of not-being**). But what does she mean by the phrase "are for thinking" in this context? Clearly she does not mean that the roads themselves, rather than those who are searching along them, are doing the thinking; so while ὁδοί ("roads") must be the subject of εἰσι ("are"), it cannot also be the subject of the active infinitive νοῆσαι ("thinking"). This appears to be why A. H. Coxon, and so many others before and after him, take it for granted that ὁδοί must rather be the *object* of the infinitive. On this reading of the syntax, the phrase "are for thinking" really means something more like "can be thought of."[4]

Once we accept this reading, however, it becomes difficult to resist the two-road interpretation. For this interpretation is plainly entailed by the following two claims, both of which are strongly supported by Coxon's reading:

> The goddess thinks that the road of being and the road of not-being are the only ones that can be thought of. (**the unconditional claim**)

> If the goddess thinks that the road of being and the road of not-being are the only ones that can be thought of, then she thinks that they are the only ones there are. (**the conditional claim**)

The second of these claims, in particular, strikes me as essentially undeniable: no one could coherently think that there is a road of being and not-being, but that this road cannot be thought of.[5] So if we are going to assume that the goddess is thinking coherently—as of course we should—then I believe we must accept the conditional claim.

Matters are somewhat less clear for the unconditional claim. That is because the goddess never uses the term δύο ("two") at 2.2, and so never explicitly says that the road of being and the road of not-being are "the only [ones] that are for thinking." Yes, they are the only ones she mentions in the bounds of Fragment 2; but suppose we were to learn—from a newly discovered papyrus, say—that she goes on to introduce a third road in the lines that directly follow 2.8, and that she does this by simply extending the syntactic formula she uses to introduce the first and second ones at lines 2.3 and 2.5: ἡ μὲν ("the one"), . . . ἡ δ[ὲ] ("another"), . . . ἡ δ[ὲ] ("and another"). In that case we would be forced to conclude that the third road also falls under the scope of the "only" at 2.2. Thus it is at least conceivable that, as F. M. Cornford in fact maintains, *she does* introduce a third road in those lines, and *it does* fall under the scope of the "only" at 2.2.[6] If Cornford is right about this, then the unconditional claim is false.

But I doubt that he is right about this. For when the goddess offers her own brief recapitulation of the argument of Fragment 2 later, in Fragment 8, she implies that the primary purpose of that earlier argument was to force a choice between exactly two roads:

> . . . And the decision about these things is in this:
> it is or it is not. But it has been decided, as is necessary,
> to leave the one [road] unthought and unnamed, for it is
>   not a true
> road, so that the other [road] is, and is genuine. (8.15–18)

If Cornford were right, then we would expect her to make some mention of a third road in or directly after these lines; but she does not. So it is hard to see how we could reasonably deny the unconditional claim on the basis of Cornford's speculation alone.[7] Given Coxon's reading of 2.2b, then, I believe we are compelled to accept both the conditional claim and the unconditional claim, and thus also whatever follows from them, including the two-road interpretation.[8]

The Way Out

But Coxon's reading strikes me as optional at best. Indeed, in at least one important respect, it is plainly inferior to an alternative reading proposed by Alexander Mourelatos almost 15 years earlier, in *The Route of Parmenides*.⁹ There he calls our attention to Empedocles's Fragment 3, and in particular to the clause I have underlined in the translation below:

> But come and consider, by every device, how each thing is manifest,
> not holding more trust in any vision than in what you hear,
> nor the echoes of hearing over the discernments of the tongue,
> and do not hold back trust in any of the other limbs,
>   wherever <u>a pathway is for thinking</u> (πόρος ἐστι νοῆσαι), but
> think each thing (νόει . . . ἕκαστον) in whatever way it is manifest.¹⁰ (9–13)

The language of this passage resonates so powerfully with that of Parmenides's Fragments 2 and 7 that I doubt it can be reasonably interpreted as anything other than a direct reply to the goddess's remarks there. So it is, by any fair standard, a more promising parallel than the ones provided by Coxon. More importantly, however, it disconfirms Coxon's reading. For although the subject of "is" at the end of line 12 is πόρος ("a pathway"), the message here is clearly not that such a pathway *can be thought about*, but that such a pathway *is potentially suitable for purpose of thinking*.¹¹ Taking Empedocles as our guide, then, we arrive at a reading of the syntax of 2.2b that allows us to deny the unconditional claim, so long as we are prepared to accept the following, revised version of it instead:

> The goddess thinks that the road of being and the road of not-being are the only ones that are potentially suitable for the purpose of thinking.

Now one concern we might have about this claim right away is its apparent implication that, according to the goddess, *the road of not-being* is potentially suitable for the purpose of thinking. Since she describes this road as "utterly uninformative" at 2.6, and denies that it is a "true" or "genuine" road at 8.15–18, we might be inclined to reject this claim out of hand. But that would be an overreaction, I think. For while it is certainly true that the road of not-being, as she sees it, does not meet all the relevant requirements for the sort of suitability at issue here, she might well hold that it *does* meet *an especially salient* one: the very same one that the road of being, as she formulates it here, also meets. Then her point would be, in effect, that the road of not-being shares an important suitable-making feature with the road of being, even though (as she quickly points out) it does not share all of them.[12]

Suppose, then, that the revised version of the unconditional claim is true. In that case the initially tempting argument for the two-road interpretation, which I laid out in the previous section, will hinge on the plausibility of a similarly revised version of the *conditional* claim:

> If the goddess thinks that the road of being and the road of not-being are the only ones that are potentially suitable for the purpose of thinking, then she thinks that they are the only ones there are.

But this claim is not even remotely compelling. For the goddess might well think that there is a road of being and not-being, but that it, unlike the road of being and the road of not-being, fails to meet the relevant requirement for suitability. So all we would need to do, if we wanted to defuse this line of argument, would be to give a reasonable account of what she thinks this requirement is, and why she thinks that the third road—unlike each of the first two—fails to meet it.

Two Accounts

There are two very different accounts we might want to give of this, however, and I believe the choice between them is a momentous one.[13] The more familiar and more traditional of the two would go something like this:

> The goddess thinks that the road of being and not-being is unsuitable for the purpose of thinking because it orients the thinker toward things that violate the law of non-contradiction.

To see more concretely how this account would work, suppose that the phrases "it is" and "it is not"—as the goddess uses them at lines 2.3 and 2.5, respectively—are to be understood as meaning "it exists" and "it does not exist." Then maybe the goddess's point is that the third road is defective because it, unlike each of the first two, orients the thinker toward things that *both* exist *and do not* exist; and it is logically impossible for there to be any such things.[14] Since this account turns on the idea that the problem with the third road is that it is somehow implicated in the violation of a logical law, I will call it **the logic-based account**.

From what I can tell, most three-road interpreters either adopt, or would adopt,[15] some version of the logic-based account. But there is an alternative account available to us here as well, thanks in large part to Kurt von Fritz's groundbreaking work on the use of noetic language (noos, noein, noēsai, noēsis, noēma) by the early Greek poets and philosophers.[16] In the passages von Fritz takes to be paradigmatic—such as *Iliad* 3.396, where Helen suddenly ἐνόησε ("apprehended") Aphrodite in the guise of an old woman—he sees a striking pattern: "The [thinker's] first recognition or classification turns out to be deceptive and has to be replaced by another and truer recognition which, so to speak, penetrates below the visible surface to the real essence of the contemplated object and at the same time, by means of this new and corrected recognition, reveals a situation of great emotional importance."[17] According to von Fritz, then, paradigmatically successful noetic activity is not just a matter of engaging in the kind of everyday thinking we associate with believing, imagining, wishing, planning, supposing, and the like; it is a far greater intellectual accomplishment than that, closer to what we might characterize as *apprehending the nature of a thing*. As he himself puts it, "the primary function of ['thinking' is] to be in direct touch with ultimate reality."[18]

Suppose he is right about this.[19] Then it is not hard to see how we might develop a new and very different account of why the goddess finds the third road deficient. Here, very roughly, is how it would go:

> The goddess thinks that the road of being and not-being is unsuitable for the purpose of thinking because it orients the thinker toward things that are not constituents of fundamental reality.

Defenders of this alternative account will be eager to point out that the full phrases the goddess uses in connection with the two roads she mentions in Fragment 2 are not in fact the contradictory pair "it is" and "it is not," but rather the manifestly noncontradictory pair "it is and it is not possible for it not to be" and "it is not and it is right for it not to be." So maybe her point here is that, while the road of being and the road of not-being orient the thinker toward things that *invariably* are and *invariably* are not, respectively, the road of being and not-being orients the thinker toward things that *neither* invariably are *nor* invariably are not: they are at some times but not others, in some places but not others, in some possibilities but not others, and so on. The difficulty she sees here, on this view, is not that there cannot be any such things, because of course there can be; it is that such things are (as we might put it) *ontologically unstable*, and for that reason cannot be constituents of fundamental reality. Since this account turns on the idea that the problem with the third road is that it fails to orient the thinker toward a privileged domain of entities, I will call it **the ontology-based account**.[20]

If my argument so far is sound, then anyone who wishes to adopt the three-road interpretation must choose exactly one of these two accounts of the unsuitability of the third road: logic-based or ontology-based. My own current view, for whatever it is worth, is that the latter is the more promising of the two, and in the concluding section I will sketch out my main reason for thinking this. But my primary aim in what follows will not be to argue that we should choose either of these two accounts; it will be to argue that we *must* choose *one* of them, because the balance of the remaining textual evidence weighs decisively against the two-road interpretation.

Diels's Conjecture

My argument takes its bearings from a familiar debate about how to fill in the lacuna in our manuscripts at 6.3.[21] Here is one possible translation of the whole of Fragment 6, with the lacuna marked by angle brackets:

> It is right both to say and to think that what-is is: for it can be,
> but nothing is not. These things I bid you to ponder.
> For < . . . > from this first road of searching,
> and then from that one, on which mortals, knowing nothing,

> wander, two-headed. For helplessness in their [5]
> chests steers their wandering thought, and they are carried
>   along,
> both deaf and blind alike, dazed, indecisive hordes,
> by whom to be and not to be are considered the same
> and not the same; and the path of all is backward-turning.

There are many challenges we face in our attempt to make sense of this passage, but only one of them requires our immediate attention here, and that is the question of how to determine the force of the missing verb in line 3. Before we try to answer this question, however, let us take a moment to pin down a few standards of adequacy for any answer we might want to give. First, a terminological matter: it is manifest in the language of 6.3–4 ("from this first road . . . and then from that one") that the goddess is discussing one road at 6.3, and then another, distinct road at 6.4–9. In what follows I will call the first one **the initial road**, and the second one **the road of wandering**. Thus the following claim should be completely uncontroversial:

> The goddess thinks that the initial road is distinct from the road of wandering.

And I take it that the placement of the lacuna, together with the same language ("from this first road . . . and then from that one") makes the following claim similarly uncontroversial:

> Whatever the goddess says about the initial road at 6.3, she also says about the road of wandering at 6.4.

So the question we are raising here could be understood more simply as follows: What does the goddess say about each of these roads—the initial road and the road of wandering—at 6.3–4?

One potentially promising place to look for an answer to this question is Fragment 7, where she expresses a clear attitude toward a certain road that looks, on the face of it, very much like the road of wandering. Here is a possible translation of the first five lines of that passage:

> For in no way may this prevail, that things that are not are.
> But you, restrain your thought from this road of searching,

and do not let habit, rich in experience, force you along this
   road,
to direct an aimless eye and echoing ear
and tongue . . . [5]

There are some obvious parallels between the goddess's description of whatever road she is discussing here in Fragment 7—which I will call **the road of habit**—and her description of the road of wandering in Fragment 6. In both passages, for example, she suggests that at least some of the mortals who search along the road she is discussing there have malfunctioning sense organs. These parallels make it safe to assume, in my view, that what she gives us at 6.4–9 and 7.2–5 are not two descriptions of two different roads, but two descriptions of one and the same road. In other words, the following claim strikes me as unobjectionable:

The goddess thinks that the road of wandering is identical to the road of habit.

Moreover, it can be shown rather easily that defenders of the two-road interpretation have no choice but to accept this claim. For it is absolutely clear from the language of Fragments 2, 6, 7, and 8 that the goddess harbors a certain negative attitude toward the road of wandering and the road of habit that she *does not* harbor toward the road of *being*. So if the two-road interpretation is correct, then she cannot think that the road of wandering and the road of habit are distinct. If she did, she would have to acknowledge at least three roads: the road of being, the road of wandering, and the road of habit. Thus the two-road interpretation straightforwardly entails that, according to the goddess, the road of wandering is identical to the road of habit.

But this claim, together with the obvious fact that the goddess forbids the young man to search along the road of habit, likewise entails that she forbids him to search along the road of *wandering*. This is an important result, given our purposes here, because it supports at least part of an answer to the question I asked at the beginning of this section: What does the goddess say about the road of wandering at 6.3–4? Given that she forbids the young man to search along that road here at 7.2, it is reasonable to anticipate that she would do the same thing there at 6.3–4. But does she? Here, again, is the crucial line from Fragment 7:

> ἀλλὰ σὺ τῆσδ' ἀφ' ὁδοῦ διζήσιος εἶργε νόημα . . .
>
> [But you, restrain your thought from this road of
>   searching . . .] (7.2)

Now notice that this line bears a marked resemblance, both semantically and syntactically, to 6.3:

> πρώτης γάρ [σ'/ τ']²² ἀφ' ὁδοῦ ταύτης διζήσιος < . . . > . . .
>
> [For < . . . > from this first road of searching . . .]

Since the explicit subject of the verb in the sentence that directly precedes this line is ἐγώ ("I"), we might expect the missing verb at the end of this line to be in the first-person singular. So it is remarkable to discover that the verb in the second person imperative at 7.2 (εἶργε, "restrain") has three first person singular forms—the present εἴργω, the imperfect εἶργον, and the future εἴρξω—that fit perfectly into the lacuna at 6.3.²³ The first of these would give us the following translation of 6.3–4a:

> For I restrain you²⁴ from this first road of searching [the
>   initial road],
> and then also from that one [the road of wandering] . . .

Because Hermann Diels seems to have been the first interpreter to propose this answer to our current question, I will refer to it as **Diels's conjecture**. Given how naturally it emerges from the considerations highlighted in this section, we should not be surprised to learn that most interpreters after Diels have accepted it.²⁵ But if it is correct, then so is the three-road interpretation. To see why, notice that Diels's conjecture immediately yields the following conclusion:

> In Fragment 6 the goddess first forbids the young man to
> search along the initial road, and then forbids him to search
> along the road of wandering.

Since we can safely assume that the goddess would *not* forbid the young man to search along the road of *being*,²⁶ we are now in a position to infer, from Diels's conjecture, that in her view the initial road is distinct not only from the road of wandering but also from the road of being. As I noted earlier, however, it is uncontroversial that in her view the road of being is distinct from the road of wandering. So if we adopt Diels's conjecture, we end up, inevitably, with three roads: the road of being, the initial road, and the road of wandering.

Resistance to Diels's Conjecture

Over the past few decades, however, many interpreters have become convinced that we have sufficient reason to reject Diels's conjecture no matter where we stand on the question of how many roads the goddess thinks there are.[27] Their view, roughly put, is that the goddess could not be forbidding the young man to search along the initial road at 6.3, because the only road she mentions at 6.1-2 that could serve as the referent of "this" (ταύτης) in the phrase "from this first road" at 6.3 is *the road of being*. Here, again, is 6.1-2a:

χρὴ τὸ λέγειν τὸ νοεῖν τ' ἐὸν ἔμμεναι· ἔστι γὰρ εἶναι,
μηδὲν δ' οὐκ ἔστιν.

[It is right both to say and to think that what-is is: for it
    can be,
but nothing is not.]

As these interpreters see it, the goddess's formulation here of what "it is right both to say and to think"—namely, "that what-is is: for it can be, but nothing is not"—closely mirrors her formulation of the road of being in Fragment 2: "that it is and that it is not possible for it not to be." This, together with her conspicuous application of the term "first" (πρώτης) to the initial road at 6.3, clearly indicates, they believe, that she identifies this road with the road of being. If their argument here is sound, then of course Diels's conjecture cannot be sustained. For, as I have noted many times before, the goddess would not forbid the young man to search along the road of being.

But I do not find this line of argument all that convincing. For there is a perfectly reasonable alternative reading of 6.1-2a that does not undermine Diels's conjecture in the slightest. Certainly we must hold that the goddess mentions the road of being in line 1, by saying that "what-is is." But I fail to see why we cannot hold, for similar reasons, that she mentions the road of not-being in line 2, by saying that "nothing is not." Many interpreters seem to assume that the only feasible way for her to mention the road of not-being here in line 2 would be for her to, in effect, *restate* it here.[28] This is a highly dubious assumption, however. All we really find in these lines, whichever interpretation we adopt, is a claim about "what-is" at 6.1 and a claim about "nothing" at 6.2a. So it is far from obvious that we find a restatement of *either* road here; on the contrary, what we seem to find, on first inspection, is a claim about the things toward which the road of being and the road of not-being, respectively, orient the thinker: *what-is*, on the one hand, and *nothing*, on the other.[29]

One might worry that this alternative reading of 6.1–2a does not really solve the problem, because the goddess would not use the term "first" in the phrase "from this first road" at 6.3 unless she were referring to the first of the two roads mentioned in the preceding lines, which (as I conceded in the previous paragraph) must be the road of being.[30] But I do not think this should trouble us very much either. For while it is certainly true that, by using the term "first" here, the goddess *could* be referring to the order in which she has just presented the road of being (first) and the road of not-being (next), she could just as easily be referring to the order in which she *will restrain the young man from* the road of not-being (first) and the road of wandering (next).[31] Indeed, on the reading of 6.1–2a I offered in the previous paragraph, the latter interpretation is far more likely to be correct; for the goddess would presumably have used ἐκείνης, rather than ταύτης, if she had intended to refer to the road mentioned less recently (the road of being, at 6.1) rather than the one mentioned more recently (the road of not-being, at 6.2a).[32] On balance, then, the case against Diels's conjecture strikes me as inconclusive.

Alternatives to Diels's Conjecture

But suppose that I am wrong about this, and that we should now be inclined to look for some alternative conjecture that does not carry the same forbidding force as εἴργω ("I restrain"). What could that verb be?

According to Cordero and Nehamas, it is a form of ἄρχω: either Cordero's ἄρξει, the second person future middle, or Nehamas's ἄρξω, the first person future active.[33] On either of these proposals, the goddess does not restrain the young man from any road at all at 6.3; rather she announces that he (according to Cordero) or she (according to Nehamas) will first "begin from" the initial road, and then "begin from" the road of wandering.

As far as I know, these are the only alternatives to Diels's conjecture that have ever been defended in print. But each of them carries significant costs, beyond the obvious oddity of suggesting that he or she (or anyone else) could begin anything more than once.[34] Indeed I believe that Nehamas's conjecture, in particular, can be ruled out on philological grounds alone.[35] For if the goddess were using some form of ἄρχω to mean "I will begin," then it would be in the middle voice, not the active.[36] In recognition of this, perhaps, Nehamas has more recently changed course, claiming that the goddess is using ἄρξω with the preposition ἀπό to mean not "I will begin [for you] from [this road]" but "I will *lead* [you] *through* [this road]."[37] As Mourelatos has pointed out, however, ἀπό cannot comfortably bear the meaning "through" with the active ἄρξω;[38] the proper translation of this phrase would have to be something like "I will lead [you] *away from* [this road]," and that defeats the purpose: Nehamas's conjecture would then have roughly the same upshot as Diels's.

Though we could avoid some of these difficulties by adopting Cordero's conjecture instead, we would still run into problems, I think. One of them is that, if we follow Cordero's advice and supply the second person form of ἄρχω, then we must abandon the better attested reading γάρ σ[ε/οι] at 6.3 in favor of γάρ τ[ε]. This is awkward, given the context, because when τε is coupled with other particles (like γάρ) it is nearly always used as a device of abstraction: a tool for drawing the hearer's attention away from the actual, concrete particularity of whatever thing, fact, or action is at issue, and toward its more general, universal, or conditional character.[39] But there is no call for this sort of abstraction at 6.3; even if Cordero were right that the missing verb is ἄρξει, the goddess would be making a very specific claim here, about a very specific person, in relation to a very specific event. Thus the decision to use γάρ τ[ε] would make little or no sense.

Moreover, and maybe more importantly, Cordero's conjecture cannot easily accommodate the textual evidence that first pointed us in the direction of Diels's. After all, it is uncontroversial that the goddess forbids the young man to search along the road of habit in Fragment 7; and, as we have seen, all parties to the debate—but especially two-road interpreters like Cordero—should accept that in her view the road of habit and the road of wandering are the same. Therefore, if Cordero's conjecture is correct, then in Fragment 6 the goddess implicitly encourages the young man to do the very same thing that, in Fragment 7, she *explicitly forbids* him to do. This is a peculiar result, to say the least, and I think we should try to avoid it if we can.

With or Without Diels

But again, suppose that I am wrong about all of this, and that we should definitely abandon Diels's conjecture in favor of either Cordero's or Nehamas's. Does it follow from this that we should also adopt the two-road interpretation, as Cordero and Nehamas themselves do? Surprisingly, perhaps, it does not. For while their conjectures do entail that the road the goddess mentions at 6.3 (the initial road) is the road of being, they *do not* entail that the road she mentions at 6.4 (the road of wandering) is the road of *not*-being; in other words, there is nothing in their conjectures, taken by themselves, that would rule out the possibility that the road of wandering is the third road (the road of being and not-being) rather than the second one (the road of not-being). So even if we decided to abandon Diels's conjecture in favor of either Cordero's or Nehamas's, we would still be perfectly free to retain the three-road interpretation.[40]

And we would still have good reason to do so, in my view. For the two-road interpretation plainly entails that, according to the goddess, the road of wandering is identical not only to the road of habit, but also to the road of not-being.[41] Having spent some time with Fragments 2, 6, and 7, however, we should find this outcome rather strange. After all, her characterization of the road of not-being is strikingly different from her characterization of the road of wandering and the road of habit.[42] At 2.5, for example, she clearly suggests that (in some sense) the road of not-being *excludes all being*, whereas at 6.8–9 and 7.1 she just as clearly suggests that (in roughly the same sense) the road of wandering and the road of habit *do not* exclude all being: they *include both* being *and* not-being. She also suggests at 2.6–8 and 8.16–8 that the road of not-being is a total nonstarter: even if you could search along it at all, which is unclear, there would be no point in doing so, since it is "utterly uninformative" (2.6). Assuming that either Cordero's or Nehamas's conjecture is correct, however, the goddess announces in Fragment 6 that either she (according to Nehamas) or the young man (according to Cordero) *will search* along the road of wandering. But if she believes that this road is a total nonstarter, then I do not see how she could think it worthwhile (or even possible) for either of them to search along it. For these reasons I cannot believe, given the evidence of Fragments 2, 6, and 7, that the goddess thinks the road of wandering is identical to the road of not-being. To that same extent, then, I cannot believe that the two-road interpretation is correct, even on the (false) assumption that we must abandon Diels's conjecture in favor of either Cordero's or Nehamas's.

The Better Path

Now that we have worked through all the relevant textual evidence with some care, I think we should be strongly inclined to reject the two-road interpretation. For, as we have seen, this interpretation turns out to be flatly inconsistent with the simplest and most straightforward proposal for filling in the lacuna at 6.3; and the only two serviceable alternative proposals—both of which are undermotivated and inferior in various crucial respects—turn out to be *consistent* with the *denial* of this interpretation. Moreover, and perhaps most strikingly, there are powerful reasons to reject this interpretation *no matter which* proposal we adopt. So at the moment, anyway, the two-road interpretation looks like a very bad bet.

But if we are going to adopt the three-road interpretation instead, then—as I argued above—we must first accept Mourelatos's (independently superior) reading of the syntax of 2.2b, and then find a way to defend the following claim:

> Though the goddess thinks that the road of being and the road of not-being are the only ones that are potentially suitable for the purpose of thinking, she also thinks that there is a third road—the road of being and not-being—that *is not* potentially suitable for the purpose of thinking.

As I said earlier, we will be able to defend this claim effectively only if we can give some account of what the relevant standard of suitability is, and why the third road, unlike either of the first two, fails to meet it. Exactly two accounts are available to us, I have been assuming: the logic-based one (the third road is unsuitable because it orients the thinker toward things that violate the law of noncontradiction) and the ontology-based one (the third road is unsuitable because it orients the thinker toward things that are ontologically unstable, and thus are not constituents of fundamental reality). Though I believe there is something to be said for each of these accounts, I close with brief explanation of why I am convinced that the ontology-based one is the more promising of the two.

Over the past twenty years or so it has become increasingly clear that certain historically underappreciated parts of the poem—the ones dedicated to cosmology in general, and to astronomy in particular—contain important, even revolutionary contributions to the development of early Greek science.[43] This should not surprise us, perhaps, given what the goddess says to the young man at the beginning of their conversation together:

> It is right for you to be told all things:
> both the unshaken heart of faithful truth,
> and the beliefs of mortals, in which there is no true trust.
> But nevertheless you will learn these things too, how it was
>   right for the believed things
> to believably be . . . (1.28–32; cf. 8.50–52)

Now I take it that these remarks at least suggest that the young man can, by searching along some road other than the road of being, learn some things that are very much worth learning. This suggestion is only reinforced and sharpened by her remarks in Fragment 10, where she seems to make another, more detailed prediction about what he can expect to learn if he searches along that other road:

> You will know (εἴσηι) the nature of the sky and all the signs
>   in the sky
> and the destructive deeds of the shining sun's pure
> torch, and whence they came to be,
> and you will be told (πεύσηι) the roving deeds of the round-
>   faced moon
> and its nature, and you will know (εἰδήσεις) also the heaven
>   all around, [5]
> from what it emerged and how Necessity led and shackled it
> to hold the limits of the stars.

Taken together with 1.28–32, this passage leaves the rather strong impression that, at some point in his discourse with the goddess, the young man will acquire robust empirical knowledge of things like the moon, the sun, and the stars. On the other hand, the argument of Fragment 8 makes it quite clear that the road of being will not orient him toward things like that; it will orient him only toward things that are ungenerated, indivisible, and unchanging.[44] Assuming, then, that the young man can acquire knowledge of a given thing only by searching along a road that orients him toward that thing, and assuming that he cannot acquire any knowledge of anything by searching along the "utterly uninformative" road of not-being, we get the conclusion that, according to the goddess, the young man will acquire knowledge of the moon, the sun, and the stars by searching along *the third* road: the road of being and not-being.

If this is true, however, then the logic-first account ends up being deeply unattractive. For it holds that the third road orients the thinker toward things that violate the law of noncontradiction; and since there can be no such things, there can be no robust empirical knowledge of them either. The ontology-first account, by contrast, does not have this implication; it merely holds that the third road orients the thinker toward things that are ontologically unstable: things that neither invariably are nor invariably are not. Since there is no in-principle obstacle to the possibility of acquiring robust empirical knowledge of such things, this account—unlike the logic-first one—can readily accommodate the assumption that the young man will, by searching along the third road, acquire robust empirical knowledge of things like the moon, sun, and stars.

Of course I do not take myself to have established here that the ontology-based version of the three-road interpretation is correct, let alone that it could be reasonably defended against the many challenges that might be, and surely would be, raised against it. All I want to insist on here is that, if it *could* be reasonably defended against these challenges, then it alone would allow us to make excellent sense of all the textual evidence we have considered so far. That is why I now believe that it, unlike the two-road interpretation or the logic-based version of the three-road one, holds out real hope for an integrated and fully satisfying reading of the poem as a whole.[45]

Notes

1. Though it is easy to forget, we should try to remember that Parmenides, like Plato, does not write in a voice that is clearly his own. Instead he casts as his narrator a certain κοῦρος ("young man") who, in the opening part of the poem,

tells of a harrowing ride in a horse-drawn chariot to the home of an unnamed goddess far from the everyday world of human affairs. With the exception of the first twenty-three lines, where the young man recounts the ride itself, the entire poem (as far as we know) is simply his report of what the goddess says to him upon arrival. In recognition of this, I will treat her, rather than Parmenides, as the primary speaker after those first twenty-three lines.

2. See, in particular, Cordero 1979, 21–24; Nehamas 1981, 97–111; and Curd 2004, 51–63.

3. Translations in this chapter are drawn, with some modifications, from Curd 2011, 55–65.

4. Though Coxon is not the only interpreter who accepts this reading, he is the only one who takes the time to defend it in much detail. He offers several parallels for his proposed construction, including an especially illuminating one from Aeschylus, where a messenger brings news to Queen Atossa of her son's catastrophic sea battle: "The sea could no longer be seen (οὐκετ' ἦν ἰδεῖν), full as it was of wreckage and the gore of mortals" (*Persians*, 419–20; Coxon 2009, 290). Other interpreters who endorse some version of Coxon's reading include Cornford 1933, 98; Guthrie 1965, 13; Barnes 1982, 159; Kirk et al. 1983, 245; and Wedin 2014, 9.

5. Here I am assuming that, if you are thinking coherently, and you think that there is a road of being and not-being, then you do not also think that this road cannot be thought of. Some interpreters might be inclined to challenge this assumption, however. Wedin, for example, seems to believe that the phrase "the only roads that can be thought of" means something along the lines of "the only roads that present themselves to the mind a priori." See Wedin 2014, 61–62. If this interpretation of νοῆσαι is correct, then my assumption is dubious; for it is not obviously incoherent to think that the road of being and not-being, unlike the road of being and the road of not-being, does not present itself to the mind a priori. But Wedin offers no parallels in support of his rather idiosyncratic interpretation of νοῆσαι, so I am not sure what to make of his proposal.

6. See Cornford 1933, 99.

7. On this point I concur with Palmer (2009, 64–65). There is a separate but important question in the background here about whether the goddess consistently maintains that *there is* a road of not-being. For she implies at 8.15–18 that it, unlike the road of being, is not a "true" (8.17) or "genuine" (8.18) road. So if she assumes that every road is a "true" or "genuine" road, as we might be inclined to do, then at 8.15–18 she baldly contradicts her earlier claims, in Fragment 2, that the road of not-being is among the "roads that are for thinking," that it is "utterly uninformative," and so on. But must we grant that she assumes this? I doubt it. Maybe she assumes instead that the standard of "truth" or "genuineness" for a road is higher than the standard of road-hood itself, so that something could satisfy the first without satisfying the second. This would be the case, for

example, if she held that a road is "true" or "genuine" only if it is not, like the road of not-being, utterly uninformative; then she could consistently maintain that the road of not-being, though it is a road, is not a "true" or "genuine" road. For an attempt to develop the contrary idea that the goddess thinks there is no road of not-being, see Smith 2020.

8. That Coxon's reading apparently entails the two-road interpretation goes either unnoticed or unaddressed by several prominent scholars who seem to accept the former but not the latter. See, for example, Barnes 1982, 157–65; Kirk et al. 1983, 245–48; and Coxon 2009, 300.

9. See Mourelatos 2008, 55–56 n. 26. He is followed here by Palmer 2009, 70–71.

10. This translation is drawn, with some modifications, from Inwood 2001, 217–19.

11. Thus the phrase "is for thinking," as it is used here, should be interpreted on the model of the phrase "were for escaping" in the following passage from Homer: "his horses were not nearby for escaping (ἐγγὺς ἔσαν προφυγεῖν), because he . . . had his attendant hold them far away" (*Il.* 11.339–41). The implication of this claim is clearly not that, had he kept his horses close, he could have escaped *them*; it is that he could have *used* them to escape *something else*. Another example from the *Iliad* is the phrase χεῖρες ἀμύνειν εἰσὶ καὶ ἡμῖν ("we too have hands for fighting") at *Il.* 13.814. See also *Il.* 23.427.

12. Compare Palmer 2009, 102.

13. Maybe there are more than two; I am not sure. But up to now I have not been able to work out any clear third alternative. So in what follows I will simply assume that the two accounts I will be considering in this section are the only ones available to us.

14. Nothing significant should hang on the supposition that "it is" and "it is not" are to be understood as meaning "it exists" and "it does not exist." The same basic story can be told on the alternative supposition that they are to be understood as meaning "it is the case" and "it is not the case," or "it is F" and "it is not F," or "it = X" and "it ≠ X," and so on. The crucial point is that the second of the two expressions is to be understood as the negation of the first, so that—according to (some version of) the law of noncontradiction—it is logically impossible for anything to satisfy both.

15. My qualification here is an attempt to abstract away from the worry, which I raised in note 8 above, that some three-road interpreters seem not to notice the apparent incompatibility between their view about how many roads the goddess thinks there are and their endorsement of Coxon's reading of the syntax of 2.2b.

16. See, in particular, von Fritz 1943, 79–93 and 1945, 223–42.

17. Von Fritz 1943, 89.

18. Von Fritz 1945, 241.

19. Some interpreters might be inclined to resist any such supposition. See, in particular, Barnes 1982, 158–59 and Lesher 1992, 103–4. But I believe their resistance is at least somewhat misdirected. Their basic point, which I take to be correct, is that noetic language is often used—by Homer, Hesiod, and even Parmenides himself—in cases where the relevant thinker is clearly not apprehending the nature of anything: he or she is merely daydreaming, wishing, deliberating, or even misjudging. This is why I doubt it would be reasonable for us to follow Kahn, Mourelatos, Coxon, and others in translating the goddess's use of νοῆσαι at 2.2b, for example, as "knowing," "realizing," "understanding," or any other such term of intellectual success. See Kahn 1969, 703 n. 4, Mourelatos 2008, 70, and Coxon 2009, 290. But von Fritz himself, as I understand him, would not disagree; his claim is not that noetic language is *always and only* used in cases where the relevant thinker is apprehending the nature of a thing, but that it is *centrally or focally* used in cases where the relevant thinker is *engaging in the sort of intellectual activity whose success consists in* apprehending the nature of a thing. So the real target of Barnes's and Lesher's critiques, I take it, is not von Fritz himself, but interpreters such as Kahn, Mourelatos, and Coxon, who have been inspired by von Fritz's work to adopt a view that is significantly less nuanced than von Fritz's own. See also Paul DiRado's chapter in this volume.

20. See Palmer 2009, 71–73 for a recent defense of one possible version of this account.

21. The material in this section on the debate over how to fill in the lacuna at 6.3 overlaps with parts of a book manuscript I am preparing for publication. For a comprehensive but compact overview of this debate, see Curd 2004, 51–63.

22. Though most manuscripts have σ[ε/οι] here, a couple of them have τ[ε], and—as we will see shortly—the majority reading rules out a potentially promising interpretation of the line.

23. Thanks to Steve White for calling my attention to the third of these forms. For an interesting argument in favor of the second over the first, see Mourelatos 1999, 125–26.

24. This proposal allows us to follow the majority of the manuscripts and read σ[ε] rather than τ[ε] at line 6.3.

25. For example, Tarán 1965, 61; Mourelatos 2008, 77 n. 7; Barnes 1982, 158; Kirk et al. 1983, 247; Coxon 2009, 59; Gallop 1984, 61; McKirahan 2010, 146; Graham 2010, 215; Wedin 2014, 55.

26. The only interpreter I am aware of who would reject this assumption is Tarán (1965, 59–61). But that is only because he thinks he needs to accept that, according to the goddess, the initial road and the road of being are the same. As I try to show in the next section, however, no one really needs to accept this.

27. In addition to the authors listed in note 2 above, see, for example, Barrett 2004, 272; Miller 2006, 5; and Palmer 2009, 65–66.

28. See, for example, Tarán 1965, 59: "μηδὲν δ' οὐκ ἔστιν, to represent the [road of not-being], would have to assert 'non-Being exists' while it means

'non-Being exists not' . . ."; Stokes 1971, 112: ". . . μηδὲν δ' οὐκ ἔστιν . . . is not a statement of the [road of not-being] at all but is a denial of [it]"; and Palmer 2009, 67: ". . . it is difficult to understand just how [μηδὲν δ' οὐκ ἔστιν] is supposed to amount to a specification of [the road of not-being]'s content."

29. Here I am in general agreement with Guthrie 1965, 22, but I find his way of putting the point a bit misleading: "Having [made] the statement that 'nothing' cannot exist," he writes, "the goddess takes the opportunity to point out that *this* way (which one must take ad sensum to be the way of thinking that 'nothing' *can* exist) is the first to be avoided." This formulation at least suggests that, if the goddess mentions the road of not-being at 6.2a, then what she means there (ad sensum) is exactly the opposite of what she says there. As I have already suggested, however, I do not think we should concede this. Compare Wedin 2014, 59.

30. One might also worry that the goddess would not use the term "first" here unless she were referring to the first of the two roads, the road of being, in Fragment 2. See, e.g., Palmer 2009, 65 and Smith 2020, 283. But I do not think we can safely assume that, in the text of the original poem, the lines of Fragment 6 follow the lines of Fragment 2 so closely as to justify this worry.

31. Yes, πρώτης ("first") is an adjective and ἔπειτα ("then") is an adverb; but this does not prevent the adjective from deriving its meaning from the order of restraint, rather than the order of presentation.

32. Here I agree, once again, with Wedin 2014, 59.

33. Cordero 1979, 22–24 and Nehamas 1981, 102–8.

34. See Mourelatos 1999, 125.

35. This despite the fact that it is by far the more popular of the two, at least among Anglophone scholars. See, e.g., Curd 2004, 58; Barrett 2004, 272; Miller 2006, 5; and Palmer 2009, 67–68.

36. On this point see Cordero 2004, 121 n. 509 and Graham 2006, 156–57 n. 29. Nehamas (1981, 105) cites LSJ ἄρχω I.2 in support of his proposal, but all the uses of the verb listed there are in the middle voice, as are both of the goddess's own attested uses of it at 5.2 and 8.10.

37. Nehamas 2002, 54–55 n. 39.

38. Mourelatos 2008, xxxiii n. 30.

39. See Denniston 1996, 528–34. As a rule, the phrase γάρ τε is found only in the work of the epic poets, where it plays the role I described above in almost every case. Within the confines of the *Iliad* alone, for example, it plays this role at 1.63, 1.81, 2.481, 3.125, 4.160, 4.261, 9.406, 10.352, 12.245, 13.279, 15.197, 15.383, 17.727, 21.24, 23.590, and 24.602. Thanks to Steve White for discussion on this point.

40. In fact, that is precisely what Palmer, a prominent defender of both Nehamas's conjecture and the three-road interpretation, recommends we do. See Palmer 2009, 67–69.

41. Here I am assuming that, according to the goddess, *there is* a road of not-being. See my discussion of the alternative possibility above in note 7.

42. Compare Palmer 2009, 68–69.
43. See, especially, Mourelatos 2011, 167–90 and Graham 2013, ch. 3.
44. See 8.1–4 together with 8.22–32.
45. Thanks to Colin Smith and Steve White for extensive discussion of the material in this chapter.

Chapter 8

Revelation and Rationality in Parmenides's Fragment 7

JENNY BRYAN

Parmenides is often read as an early champion of rationality and deductive argument. Indeed, he has often been claimed to be the Western tradition's first genuine philosopher.[1] And yet, as is also often acknowledged, several aspects of his poem sit less well with his image as a rationalist. The most obvious of these is the fact that his poem presents his philosophical innovations as the content of a revelatory address by a goddess. This apparently "irrational" frame—along with the divine authority for his teachings that it implies—has been addressed in a variety of ways by those who wish to emphasize Parmenides's philosophical credentials. Often, this is a matter of minimizing the significance of the revelatory status, perhaps by reading the Proem's description of the journey to meet the goddess as allegorical or as a mere nod to poetic convention. Such readings assume the priority and authority of Parmenides's arguments and seek to read the rest of the poem as supporting, or at least not undermining, that priority.[2]

The final lines of Fragment 7 are often treated as a key text in justifying the rationalist reading of Parmenides that elevates the authority of reason and argument over the didactic or revelatory authority of the goddess. Indeed, read in this way, it is the goddess *herself* who grants permission for her audience to detach themselves from her apparently irrational authority. Although Parmenides puts all his arguments into the mouth of a goddess, Fragment 7 is taken to represent the goddess's demand that these arguments should be assessed on their own terms, thereby rendering their status as a revelation relatively insignificant.

From antiquity, readers have interpreted Fragment 7 as a fundamental statement of Parmenides's commitment to the power of reasoning:

οὐ γὰρ μήποτε τοῦτο δαμῇ εἶναι μὴ ἐόντα
ἀλλὰ σὺ τῆσδ' ἀφ' ὁδοῦ διζήσιος εἶργε νόημα
μηδέ σ' ἔθος πολύπειρον ὁδὸν κατὰ τήνδε βιάσθω,
νωμᾶν ἄσκοπον ὄμμα καὶ ἠχήεσσαν ἀκουήν
καὶ γλῶσσαν, κρῖναι δὲ λόγωι πολύδηριν ἔλεγχον
ἐξ ἐμέθεν ῥηθέντα.³

[For never at all could you master this, that things that are not are;
But as for you, keep your thought away from this road of investigation,
And do not let much-experienced habit, force you down onto this road,
To wield an aimless eye and an echoing ear
And tongue; but κρῖναι λόγωι πολύδηριν ἔλεγχον
Spoken by me.]⁴

Our earliest source for the second half of the fragment, Sextus Empiricus, introduces it as an expression of the supremacy of reason over sense perception.⁵ Modern readers have tended to follow this tradition, with justification; Fragment 7 does seem to draw a contrast between traditional modes of inquiry based on and communicated via the senses, and some alternative route of inquiry independent of the senses. In pursuing such readings, however, modern interpreters have tended to expand the significance of the instruction given by the goddess in the final lines, κρῖναι λόγωι πολύδηριν ἔλεγχον ῥηθέντα, often translated as something like "judge by reason/reasoning the controversial testing/refutation spoken by me."⁶ Two connected claims are often made on the basis of readings of 7.5–6. The first is that these lines represent an endorsement of independent reasoning as means of assessing the value of the account of reality and knowledge set out in Parmenides's poem as a whole. The second is that, as an endorsement of independent reasoning, these lines undercut or deflate the apparently irrational authority of the goddess to whom Parmenides attributes his arguments. We can see how these two claims work together. For if Fragment 7 serves to establish reason as the final arbiter of truth, the authority of the goddess as a source of revelation is presumably rendered unnecessary; the goddess may be important as the origin of the arguments, but her authority does not extend to verifying their success. So, for example, Daniel Graham suggests that, with the final lines of Fragment 7, "the goddess affirms that reasoning is the final arbiter of her account, not tradition, revelation, or divine authority. Clearly Parmenides is a genuine philosopher, despite the mythological trappings and the poetic medium of his book."[7] So, although Parmenides succumbs to the tradition of Homer and Hesiod in attributing his arguments to a revelatory goddess, Fragment 7 is taken to grant permission for philosophical readers to set the goddess aside, along with the irrational, apparently unphilosophical, authority that she represents.[8]

This way of interpreting the final lines of Fragment 7 thus provides a justification for a common deflationary attitude toward the role of the goddess within the poem as a whole. Given that all the arguments of the poem are attributed to the goddess, this kind of attitude toward her authority will also inform how we approach those arguments. Bearing all this in mind, it is worth taking the time to scrutinize such readings, and to ask whether they provide a firm foundation for broader interpretation of the poem and, in particular, whether they justify the tendency to minimize the significance of the goddess as the source of Parmenides's teachings. The value of such scrutiny becomes even clearer when we recognize that these lines are packed with vocabulary, the sense of which has been notoriously difficult to decide with any certainty. In what follows, I will argue that there is good reason to doubt that Fragment 7 represents an endorsement of independent reasoning over and above the authority of the goddess. I will suggest instead that these lines serve to reinforce the goddess's status as didactic guide for the kouros. This is not to say that there is no room for the application of independent reasoning in assessing Parmenides's philosophy. My suggestion is simply that those who want to claim that Parmenides is a "genuine philosopher" who asserts reason as the final arbiter of truth, and thereby to deflate or disregard the significance of divine authority within his poem, need to work harder to justify attributing this view to Parmenides generally, and to Fragment 7 specifically.

"Judge by Reason?"

It is particularly striking that, although the term κρῖναι—often, as I noted above, translated into English as "judge"—is fundamental to our understanding of the goddess's instruction, its meaning here has received little scrutiny.[9] Thinking carefully about the meaning of this term in particular is important because, as we shall see, it seems to have had a significant role in shaping assumptions about the broader sense of the fragment.

The verb κρίνω generally expresses acts of selection, discrimination, or choice. It occurs most frequently in Homer to describe the selection or separation of groups of men from among the troops.[10] It is also occasionally used to describe decisions made in quasi-legal circumstances, as at *Iliad* 16.384–94, which describes Zeus's punishment of those who "make crooked decisions (σκολιὰς κρίνωσι θέμιστας)."[11] Considering this Homeric background of the word, alongside Parmenides's broader use of quasi-forensic vocabulary of the type later found in forensic and deliberative rhetoric (σήματα, δίκη, πίστις and, indeed, ἔλεγχος), we can see that "select," "decide," and "judge" are all plausible candidates for translating κρῖναι in Fragment 7.[12]

In reading Fragment 7 as an endorsement of independent reasoning, scholars have tended to prefer to translate κρῖναι into English as "judge" in a sense that goes beyond a simple act of selection or decision-making.[13] In fact, the term is often taken to describe the process of examination or assessment on which such a decision might be based. So, for example, Barbara Sattler has interpreted Fragment 7's κρῖναι as combining the senses of "distinguish," "interpret," "examine," and "judge" to describe the process of *both* analyzing an argument *and* assessing the truth of its conclusion.[14] Sattler therefore understands the goddess's instruction to "judge" as an indication that we should be "checking, verifying, [and] proving her elenchos," that is, ratifying the goddess's authority through the application of rational assessment. Similar suggestions that the goddess is requesting an independent evaluation of her arguments are common.[15] This is, it seems to me, a lot to pack into the sense of a word that has a basic sense of "discriminate," "select," or "choose," but we can see why such a move is often made. After all, we are inclined to think that a well-made judgment or choice, especially within a philosophical context, certainly should be justified by careful and independent evaluation. And yet, as we will see, there is good reason to think that the goddess is not, in fact, instructing the kouros to judge what she has said in the sense of analyzing and assessing her arguments. Rather, she is instructing him to make a particular choice: a choice that reinforces rather than undermines her own authority.

Parmenides uses cognates of κρῖναι several times elsewhere in the poem. Prior to Fragment 7, the goddess refers to the "mortals who know nothing" as ἄκριτα φῦλα, "uncritical tribes," as part of Fragment 6's criticism. As the rest of the fragment explains, they are "uncritical" in the sense that they fail to distinguish properly between what-is and what-is-not, thinking them to be "both the same and not the same." Now, of course, we could say that this failure in understanding (whatever its precise nature) is the result of a lack of the analysis scholars have suggested is encouraged by Fragment 7. Nevertheless, the sense of ἄκριτα here is *primarily* one of failing to *discriminate* or *decide* between what-is and what-is-not, as opposed to failing to undertake the sort of analysis on which such an act of discrimination might be based.

This use of κρίνω as describing an act of discrimination or selection is also found in the goddess's account of mortal thinking toward the end of Fragment 8. At 8.55–56, the goddess complains that mortals

> . . . have distinguished (ἐκρίναντο) opposites in body and concocted signs
> Separate (χωρίς) from one another . . .

She goes on to complain that this distinction has no real substance, since, as Fragment 6 has already made clear, it is informed by an inability to distinguish between what-is and what-is-not. The mortal discrimination described at 8.55–6 is thus, in fact, a failure to discriminate. It is an illegitimate κρίσις (decision) undermined by the fact that those who make it are ἄκριτα (uncritical). And once again, while it is fair to suggest that this false discrimination is aligned with a failure to understand the arguments presented by the goddess against the viability of what-is-not as an object of thought or language, the goddess's explicit concern here is with the *act of discrimination* rather than with the process by which it is (or is not) justified.

The end of Fragment 8 explains that the goddess regards mortals as generally unable (or unwilling) to distinguish in the correct way between what-is and what-is-not. As a result, they make distinctions that are essentially illegitimate, understanding the world in terms of opposites such as light and dark. However, Fragment 8 also provides us with an example of a legitimate κρίσις. In lines 8.15–18, the goddess declares that

> ... the decision (κρίσις) about these things is in regard to this:
> It is or it is not; and so, it has been decided (κέκριται), as is necessary,
> To let alone the unthinkable, unnameable (for the road
> Is not real), and that the other is and is true.

Here again, we see that the sense of both the noun (κρίσις) and the verb (κέκριται) is one of selection and decision. The "decision" of line 8.15 is said to be in relation to what-is and what-is-not. Line 8.16 then tells us that a decision has already been made to "leave alone" (ἐᾶν) the route of what-is-not. Once again, in both cases, we could argue that such decisions ought to be made on the basis of a process of analysis and assessment of arguments, but once again it is not *this process* to which the goddess is explicitly referring but rather to the choice itself. This is particularly clear in the case of line 8.16, where κέκριται with the infinitive ἐᾶν indicates that the goddess is describing a *decision to act* in a particular way.

We have seen that, where cognates of κρῖναι appear elsewhere in Parmenides's poem, it is to describe an act (or failure) of selection or discrimination, uses that reflect the primary sense of the word as it occurs in Homer. Even if we were to translate these occurrences with variations of "judge" or "judgment" (translating ἄκριτα as "without judgment," e.g.),[16] something further would be required to justify imposing the idea of extended analysis or assessment in each case.[17]

Bearing this in mind, we can return to κρῖναι in Fragment 7. There is no compelling reason to assume that Parmenides is using the word differently here from the way that he uses it elsewhere in the poem, that is, to describe an act of discrimination or selection. Indeed, it seems to me that we have good reason to assume that he is using it in more or less the same way, since Fragment 7 as a whole establishes a contrast between two options. The goddess's positive encouragement in the final lines is set against the warning of lines 7.1–5 not to pursue the route of inquiry along which mortals let habit force them. In encouraging the kouros to make a choice, the goddess is exhorting him to liberate himself from the "uncritical" habits of his kind: she is urging him to make a choice that most mortals are unable or unwilling to make. Reading κρῖναι as an instruction to "discriminate" or "choose" not only aligns the sense of the word in Fragment 7 with the way it is used elsewhere in the poem, but it also connects the choice encouraged here with that which mortals are shown elsewhere to fail to make, as well as with that which the goddess says has been made at the beginning of her arguments in Fragment 8.[18] In addition, and importantly for my purposes, it maintains the authority of the goddess. For she is not inviting the kouros to verify her arguments. Rather, she is instructing him to make the distinction that will enable him to accompany her on the route toward knowledge.

It is worth emphasizing that I am not, by any means, denying the possibility that Parmenides intends his poem to encourage the independent rational assessment of the arguments it offers. The very fact that Fragment 8 presents arguments acts as an implicit invitation to consider their success. I am, however, suggesting that those who wish to read Fragment 7 as an *explicit instruction* to assess her arguments independent of her authority must work harder to justify that position, especially when such interpretations are used as the basis for framing an approach to the poem as a whole.

At this point, it will be useful to say something about Fragment 7's use of λόγος. For those seeking to read the goddess's instruction κρῖναι as a command to apply independent assessment are perhaps further encouraged by the possibility that, whatever it is that she is asking the kouros to do in these lines, she is asking him to do it "with reason" (λόγωι). Indeed, even if we agree that κρῖναι is best translated as something like "discern" or "decide," perhaps λόγος is where we can locate the process of analysis identified by many in Fragment 7. So, for example, Patricia Curd glosses Fragment 7's λόγος as an instruction "to judge by *thought, argument or reasoning*" (my emphasis).[19]

Here again, however, the sense of the term cannot be assumed with any certainty. As is often acknowledged, it is unclear whether logos has already attained the sense of "reason" or "reasoning" by the time Parmenides is composing. If so, then Fragment 7 would represent the first extant instance of it carrying this sense.[20] More significantly, it would be the only instance in Parmenides in which he uses the term to mean "reason" as opposed to "account" or "discourse."[21] For this reason, many scholars persist in translating logos here as something like "discourse" or "argument," but then struggle to give an account of what it means to "judge" (in the sense of developing an extended analysis) "in discourse."[22]

Here again, I suggest, the assumption that the goddess is explicitly endorsing a process of analysis or assessment is at fault. If we understand that the goddess is instructing the kouros to make a choice, we can see how that choice can, in a sense, be made "by" or "with" or "in discourse." Looking again at Fragment 7 as a whole, we see that the goddess warns the kouros to avoid the illegitimate route of inquiry pursued by his confused fellow mortals. In fact, she explicitly warns him not to "let much-experienced habit force you down onto this road,/ To wield an aimless eye and an echoing ear/ And tongue." This is, I take it, a prohibition against adopting or repeating the kind of confused accounts promulgated by mortals. Thus, Fragment 7 represents two closely connected instructions. First, the kouros must reject the incoherent route of inquiry that takes what-is-not as a viable subject matter and thereby produces the traditional, confused discourse of mortals. Second, he must follow the goddess in rejecting what-is-not, that is, in deciding against the viability of what-is-not, "in discourse" and thus speaking only of what-is.[23]

Such a reading of Fragment 7's instruction κρῖναι λόγωι is, I suggest, supported by the first line of Fragment 6: "It must be that what is there for speaking (λέγειν) and thinking is." Here, the goddess is asserting that speech, that is, λόγος, like thought, can take only what-is as its object. Fragment 6 goes on to spell out the ways in which mortals traditionally fail to recognize this restriction. On the reading I am setting out, Fragment 7 reframes this assertion as an instruction to the kouros to maintain this restriction in his own discourse (λόγωι), that is, to follow the example of the goddess.

I have been suggesting that Fragment 7 does not present an explicit endorsement of independent reason in the way that has so often been assumed. Scholars who wish to read κρῖναι as meaning "assess" or "analyze" and even λόγῳ as meaning "by reason" or "by reasoning" ought to do so on the basis of clear argument rather than assumption. The alternative reading that I have developed takes the goddess to be reinforcing her own authority, rather than setting it aside in favor of the power of independent reason and argument. The goddess is inviting agreement rather than verification.[24] The question remains, however, as to whether such a reading can also incorporate a plausible interpretation of perhaps the most controversial term in these final lines of Fragment 7: ἔλεγχος. I suggest below that there is indeed a way to understand this word that fits with my translation of κρῖναι λόγωι as 'decide' or "discriminate with [your] discourse."

"Controversial Testing"?

While the sense of κρῖναι has received relatively little attention, the meaning of Fragment 7's ἔλεγχος has prompted much detailed discussion, motivated in part by the term's significant philosophical afterlife in the Socratic tradition. James Lesher has presented a detailed study of the early development of the term, arguing that Parmenides's use is best translated as "test" or "examination."[25] Others have preferred the more explicitly negative "refutation," even in the face of Lesher's criticism of such a translation on the basis that no refutation can be found in the poem prior to Fragment 7.[26] Certainly, there is good reason to think that the term incorporates the negative sense that its cognates express in Homer, where the term is one of "shame" or "failure."[27] There is also good reason to think that it refers to something earlier in the poem, since as the goddess uses the aorist participle ῥηθέντα to qualify her reference to her ἔλεγχος.[28]

It is notable how often interpretations of Parmenides's ἔλεγχος assume that its referent must be some sort of extended analysis or argument. Having surveyed the standard interpretations of κρῖναι above, we can see what motivates such assumptions. For if κρῖναι is assumed to describe a process of analysis or assessment, the ἔλεγχος that the kouros is instructed κρῖναι must be a suitable subject of such analysis. So, for example, Sattler argues that the rejection of what-is-not in 7.1, since it is merely an assertion, cannot stand as the referent for the ἔλεγχος of line 5, because "if the goddess does provide a refutation, a test or an investigation that can be judged with logos, this line has to be an assertion of an argument that was already given before."[29] Sattler identifies this argument (which she in fact takes to be a refutation) in the final lines of Fragment 2 in combination with Fragment 3, with their rejection of what-is-not as a viable object of thought and knowledge. The question whether these texts provide a satisfactory candidate for classification as either an argument or a refutation is controversial.[30] In fact, the concern to locate the argument to which ἔλεγχος is assumed to refer, and to which the kouros is supposed to apply his independent skills of rational analysis, has prompted some to look for a way of reading the term as a reference to those arguments set out in Fragment 8, the qualification ῥηθέντα notwithstanding. Curd, for example, has argued that Fragment 7's retrospective reference to an ἔλεγχος can also be read forward concerning Fragment 8, the arguments of which certainly seem to invite the kind of extended analysis assumed to be described by κρῖναι.[31]

I have argued above that there is no compelling reason to assume that Parmenides's κρῖναι describes such a process of extended and independent analysis of an argument. Indeed, I suggested that there was good reason to read it simply as an instruction to "decide," "select," or "discriminate." If the term ἔλεγχος necessarily implies some kind of argument or disproof, then we can see that this would indeed provide evidence in favor of reading κρῖναι in the more developed, philosophical sense. In fact, however, ἔλεγχος need not carry be read as carrying this sense. Rather, as Alexander Mourelatos has suggested, Parmenides's ἔλεγχος could simply retain the Homeric sense of "shame" or "reproach."[32] In fact, even if we follow Lesher in reading it as meaning something like "test," it is not obvious that we need to assume this to imply an extended process of argument or assessment that is itself a suitable object of evaluation.

Among the early examples of ἔλεγχος cited by Lesher, we find Pindar *Nemean* 8.20–21:

> For many things have been said in many ways but, discover-
>   ing new ones, to put them to
> the touchstone (δόμεν βασάνῳ)
> For testing (ἐς ἔλεγχον), is utter peril.

Clearly, ἔλεγχος in this context describes some kind of test. It describes the process of testing a metal against a touchstone (βάσανος) by striking the former on the latter and seeing whether it leaves an obvious trace, thereby proving it to be a precious metal. This is certainly the kind of test that encourages a judgment or decision—the decision to classify the metal as precious or not—but that, in itself, requires no obvious process of analysis, interpretation or verification to reach that decision.[33]

It is, I suggest, possible to read Parmenides's ἔλεγχος in much the same way as that of *Nemean* 8, namely as describing a test against a criterion that any particular candidate will either pass or fail.[34] Considered in this way, without the need to identify an extended argument, we can read Fragment 7's ἔλεγχος as a reference to the rejection of what-is-not in Fragment 2 and the rejection of mortal discourse founded on it described in Fragments 6 and 7. On such a reading, we are no longer looking for an extended or successful argument *in favor* of that rejection, and so we need not worry whether we can identify one that we find philosophically satisfactory. By the time we reach the instruction of Fragment 7, we have been told several times that what-is-not is not a viable object of inquiry, and that the route of inquiry pursued by mortals, which assumes the existence of what-is-not, is not to be followed. I suggest that this rejection of what-is-not is the ἔλεγχος to which the goddess refers. It is an ἔλεγχος not in the sense of an argument about the viability of what-is-not held up for the independent assessment of the kouros. Rather, it is an ἔλεγχος ("test") in the sense of being an assertion of the failure of what-is-not to provide a suitable subject of thought and discourse. When the goddess instructs the kouros to κρῖναι her ἔλεγχος with λόγος, she is exhorting him to decide in favor of—and thus accept and adopt—her critique of what-is-not in the way that he speaks about reality.[35]

This suggestion that the goddess is instructing the kouros to adopt her rejection of what-is-not in his own discourse, and thereby to follow her example and thus the correct route of inquiry, has two advantages. The first is that it gives good sense to the adjective πολύδηρις as meaning either "strife-filled" or "strife-causing," or, indeed, both. When the goddess describes her ἔλεγχος as πολύδηρις, she is acknowledging the conflict resulting from her blanket rejection of mortal discourse that assumes the existence of what-is-not.[36] If the kouros follows the goddess's example in making the act of selection or discrimination that results in the exclusion of what-is-not, he too will find himself in constant conflict with his fellow, "uncritical," mortals, since everything they say is informed and undermined by their failure to reject what-is-not.

The second advantage of this reading is that it does not require us to assume any conflict between the instruction given by the goddess in Fragment 7 and the divine authority she is shown to hold throughout the poem (including and especially in Fragment 8). I have been particularly concerned to scrutinize those readings that seek to use Fragment 7 as a justification for setting aside the divine authority of the goddess, or even treating it as a mere literary device. It is worth noting, however, that the reading developed above maintains the authority of the goddess while *also* making room for the application of independent reasoning on the part of the kouros. For while I have argued that the goddess is not inviting the kouros to analyze her own arguments independently of her authority, we can read her instruction to "decide by discourse my controversial critique" as an exhortation to apply his reasoning to the accounts consistently given by his fellow mortals and, once equipped with the content of her revelation, to maintain his resistance to incoherent talk of what-is-not.[37]

Conclusion

I have presented a critique of standard readings of the final lines of Fragment 7, which interpret them as an endorsement of independent reasoning and, thus, as a resignation of authority on the part of the goddess to whom Parmenides attributes his arguments. I have suggested that such readings tend to assume too much about the possible meaning of κρῖναι in particular. For while scholars often read this as an instruction to undertake a process of independent assessment of (some part of) the goddess's account, it more naturally carries the sense of "decide," "select," or "discriminate." It is in this sense that Parmenides uses cognates of κρίναι elsewhere in his poem, and there is good reason to assume a connection between the acts of choice and discrimination she describes elsewhere and that which she instructs the kouros to undertake in Fragment 7. Likewise, although readers preoccupied with Parmenides's philosophical credentials are often keen to identify an endorsement of reasoning in the occurrence of λόγος in these lines, to read the word in such a way requires us not only to assume that this marks the first extant instance of it as carrying the sense of "reason" but also to ignore the fact that Parmenides himself uses it elsewhere to mean "account," "words," or "discourse." I have also demonstrated that even if we adopt the conclusions of Lesher's careful study of the early development of the term ἔλεγχος, and translate it as "test" or "testing," this need not imply that it describes an extended process of argument or analysis such as would be an appropriate object of assessment by the kouros. On these grounds, I have suggested that we should be wary of the assumptions that underlie translating κρῖναι λόγωι πολύδηριν ἔλεγχον/ ἐξ ἐμέθεν ῥηθέντα as "judge by reason the controversial testing/ spoken by me." I have argued that we should, instead, adopt a translation along the lines of "decide by discourse the controversial testing/ spoken by me." Rather than inviting the kouros to assess her arguments independently of the authority of their source, the goddess is reinforcing her own authority, instructing the kouros to undertake the act of discrimination of which his fellow mortals generally prove themselves incapable: the act of discrimination that the goddess goes on to declare *has been made* at 8.15–18, thereby enabling her to set out her account of what-is.

I am aware that my reading of Fragment 7 will be disappointing for many. After all, this Parmenides remains committed to the "irrational," unphilosophical authority of the goddess in Fragment 7 and beyond. He does not grant us permission to ignore her or to deflate her significance as the source of Parmenides's account of reality. This Parmenides thus does not fit the mold of a "genuine philosopher" in the modern sense of the word. I confess, however, that this objection does not worry me as much as others might think it should. For I see no reason to think that

Parmenides regarded the divine authority of the goddess as obviously in conflict with the authority of reason, or that he would have understood an appeal to divine authority as "unphilosophical," even if he possessed a developed notion of "philosophy" itself.[38]

It is worth emphasizing, that, although I do not think that Fragment 7 grants permission to ignore the role and authority of the goddess in the poem as a whole, nor do I think that my deflationary reading of the text justifies a blanket rejection of philosophical or analytical readings of the poem. As I have sought to emphasize above, while I do not agree that Fragment 7 voices an instruction to assess the arguments of the goddess on their own terms, I do think that the fact that the goddess offers arguments, particularly in the extended account of Fragment 8, marks an invitation to assess and understand them. I have also suggested that the kouros is encouraged to implement his own ability to reason in the way that he responds to and recognizes the inadequacy of traditional mortal accounts. On this reading, Parmenides makes room for the authority of both the goddess and reason.

Notes

1. See Bryan 2020, 218–22 for a more detailed discussion of modern approaches to Parmenides.

2. See Bryan 2020, 222–26.

3. Fragment 7 is a reconstruction from several sources: lines 1–2 are cited as a unit by Plato *Sophist* 237a and Simplicius in *Phys.* 143. Sextus Empiricus *Adv. Math.* 7.111 cites lines 2–6 as a unit (in conjunction with Fragment 1 and the first line of Fragment 8).

4. Translation adapted from Laks and Most 2016, 99. Since the meaning of κρῖναι λόγωι πολύδηριν ἔλεγχον is the subject of the current chapter, I leave it untranslated for now.

5. *Adv. Math.* 114. See also Diogenes Laertius 9.22.

6. Kirk, Raven, and Schofield 1983, 248, "judge by reason the strife-encompassed refutation"; Gallop 1984, 63, "judge by reasoning the very contentious disproof"; Palmer 2009, 367, "judge by reason the strife-filled critique"; Graham 2010, 215, "judge by reasoning the very contentious examination uttered by me"; Mansfeld and Primavesi 2011, 325, "Beurteile in rationale Weise die streitbare Widerlegung."

7. Graham 2010, 237. Granger (2010, 30) characterizes the goddess as "someone who respects argumentation [and] would be only too eager to invite her audience to assess the value of her arguments through their own capacity for reasoned arguments." See also Curd 2004 [1998], 20 and Sattler 2011, 33.

8. Curd (2004 [1998], 20) suggests that Parmenides is appealing to the goddess as "a source of inspiration" rather than as a justification for his philosophical claims. Sattler 2011, 32–33 argues that Fragment 7 represents an assertion that "mere authority is excluded as a criterion," suggesting that Parmenides uses the goddess to "stress the importance of his subject."

9. We can usefully compare the extent of scholarly discussions of Fragment 7's κρῖναι with those of ἔλεγχος (the meaning of which I discuss below).

10. See, for example, *Iliad* 2.362, 13.129; *Odyssey* 4.408, 9.90.

11. See also *Odyssey* 12.438–41.

12. See Bryan 2012, 80–93 on Parmenides's use of forensic vocabulary.

13. Laks and Most (2016, 99) retain the sense "decide": "by the argument, decide the much-disputed refutation." Coxon (2009, 311) originally translates κρῖναι as "decide" but adopts "judge" in his commentary on the passage.

14. Sattler 2011, 31–32.

15. See, for example, Curd (2004 [1998], 63), who reads Fragment 7 as exhorting the audience to "make a judgment not on the basis of habit or experience, *but having listened to and considered* the account that the goddess has given in the *Aletheia* and in the *Doxa*" (my emphasis). See also Mourelatos (2008, 91), who translates κρῖναι as "judge *for yourself*" (my emphasis).

16. As does Coxon (2009, 303), although note that he goes on to suggest that the sense of ἄκριτα is "clarified" by 7.5 and 8.15, translating κρῖναι as "decide" and κρίσις as "decision."

17. Lesher (1984, 17) notes that κρίνω tends to denote a decision, as distinguished from a process of deliberation. See, for example, Aeschylus *Eumenides* 443: "Test him, and make a straight judgment" (ἀλλ' ἐξέλεγχε, κρῖνε δ' εὐθεῖαν δίκην). Ebert (1983, 185–89) suggests that early uses of κρίνω generally describe acts of "deciding" (between options or about a case), rather than "judging" in the sense of "judging the worth of something."

18. See Mourelatos 2008, 312 n. 32 on the likely contrast intended between 8.55 and fr. 7.

19. Curd 2004 [1998], 62. See also Sattler 2011, 32–33, who reads logos in Fragment 7 as indicating that "κρίνειν is thus a *rational investigation, based on reason*, with the help of which we differentiate and thus proceed to a well-grounded judgment." Sattler emphasizes that it is logos (as "reason") that allows us to exercise analytic independence.

20. See Granger 2010, 31–33 for a useful survey of the scholarship.

21. See lines 1.15 and 8.50.

22. Mourelatos (2008, 91) and Barnes (1982, 170) translate logos as "argument." Tor (2017, 184–85 and 340 n. 57) shifts between "reasoning" and "account" and suggests that, whatever the correct sense of logos, it refers forward to the arguments of Fragment 8. Note that Coxon (2009, 310–12) translates κρῖναι λόγῳ as "decide by discourse," but adopts "judge by reason" in his commentary.

23. I thus read λόγος here as referring to whatever account the kouros himself produces in response to the teaching of the goddess, rather than to that of the goddess herself.

24. I am thus partly in agreement with Kingsley 2020, 129–44. I see no need to emend the dative λόγῳ to the genitive λόγου, as he proposes, if we assume that the goddess is issuing an instruction to the kouros to adopt her rejection of what-is-not in his own (future) discourse, thereby rejecting the confused language of his fellow mortals.

25. Lesher (1984) reads Parmenides's ἔλεγχος as the process of testing and rejecting different routes of inquiry.

26. See, for example, Sattler 2011, 29–31 and Granger 2010, 31 n. 39. As Granger notes, Lesher established the lack of a "refutation" only in the narrow sense of a disjunctive syllogism.

27. See Mourelatos 2008, 91 n. 46 and Palmer 2009, 109.

28. As noted by Verdenius 1942, 64.

29. Sattler 2011, 29. Lesher (1984, 17) reads ἔλεγχος as "a process of examination or a competitive testing" resulting in an act of judgment or decision.

30. See Palmer 2009, 108–9.

31. See Curd 2004 [1998], 62–63, especially n. 107. See also Tor (2017, 185), who reads Fragment 7's λόγος as referring to the arguments of Fragment 8.

32. Mourelatos 2008, 91 n. 46.

33. See also *Olympian* 4.17–18, where "experience" is said to be the "test" (ἔλεγχος) of men. This could be understood to imply an extended process of testing, but the comparison given is of victory in a footrace, that is, an event rather than a process.

34. Note that I am not denying that early instances of ἔλεγχος can describe an extended process of testing, only that it can sometimes be used to describe a straightforward pass/fail test against a simple criterion.

35. This is a less rigorous and specific instruction than that suggested by Coxon (2009, 311), who reads 7.5-6 as encouraging the kouros "to use as a test [. . .] the law of contradiction" he takes to have been set out in Fragment 2.

36. There is, contra Coxon 2009, 312, no need to pick out a particular school of natural philosophy as a target here. It is mortal discourse in general that the goddess has been criticizing in Fragments 6 and 7.

37. On this reading, then, Fragment 7 anticipates and partially explains the promise at 8.60 that "no opinion of mortals shall ever overtake you": in making the correct κρίσις as instructed in Fragment 7, the kouros is equipped to see the fundamental failures of all traditional mortal accounts of reality.

38. See Tor 2017, 10–19 for an insightful discussion of the traditional scholarly dichotomy between the rational and irrational in approaches to Parmenides.

Chapter 9

Signposts for the Study of Nature
Parmenides's Fragment 8

Eric Sanday

Fragment 8 represents a curious shift in the manner of argument employed by Parmenides in previous fragments. In Fragments 2 and 6, Parmenides has demonstrated the impossibility of nonbeing strictu sensu, thereby establishing being (ἔστι) as the necessary source for all structure determinately intelligible in terms of what each thing in its nature both "is" and "is not." That insight continues to be one of the most radically transformative moments in the history of philosophy, and it remains to this day one of its most challenging. It requires that we see and preserve the radical difference in kind between the apprehension of being and the structured relations of "is" and "is not" in which all intelligibility is discursively articulated. The specific puzzle driving Fragment 8 is to preserve the insight into being (ἔστι) in and through its discursive articulation as a being of a specific kind. To understand and preserve our own insight, ἔστι itself must be subjected to the relational terms of being and not-being, what it is and is-not, and established as something posterior to and derivative of itself. I will argue that Fragment 8 is the context in which Parmenides shows the various appropriate discursive articulations of being in light of the basic structures of things in the world over which being has causal priority. Fragment 8 is, on this reading, a transition from Truth to Doxa, in which the insight into being is preserved at various distinct levels of reception.

Once we adopt this perspective, we will be able to resolve questions that could not fail to arise for any reader of Parmenides's poem.

Key to my argument is that we differentiate between being as such and its various general determinations, including determination of natural kinds (8.6–8.21), complexity (8.22–8.33), and the activities of naming and the productive arts (8.35–8.51), which will be three foci of the main sections of this chapter. The overarching claim is that in Fragment 8 being is determined as a structuring cause relative to the basic aspects of determinate intelligibility in the world. As I will argue, Parmenides engages with a range of specific notions in the course of Fragment 8, including birth, growth, complexity, multiplicity, thinking, naming, and production in order to preserve the tension between being as such and being as structuring cause, and ultimately to arrive at a well-oriented understanding of things in the world.

To be more precise, the goddess opens Fragment 8 by characterizing the being (ἔστι) studied in Fragment 2 as what-is (τό ἐόν, the nominalized form of being), transforming being into something we can posit as the causal source responsible for the order of things in the natural world. Fragment 8 preserves the insight into being (ἔστι) as a specific kind of being, "what-is" (τό ἐόν), clarifying it in terms of the attributes that accord to some of the key ways in which we conceive of beings. Having thus shifted the terms of analysis from the sheer apprehension to the discursive grasp of being, the goddess declares that what-is is ungenerated (ἀγένητον), of a single kind (μουνογενές), steadfast (ἀτρεμές), and not ending (ἠδὲ τελεστόν), thus subjecting being to relational terms while stressing its independence from those terms. She goes on to describe what-is as uniform, evenly balanced, cohesive, and all alike, stressing structures of spatial and temporal particularity. The so-called signposts to which she refers are the images and arguments that portray what-is in terms most appropriate to it. These various accounts point to what-is in very different registers, some of which seem much less precise than others, such as likening what-is to the bulk of a well-rounded ball. Images of this sort have caused interpreters significant trouble, leading some scholars to take all attributes literally, others metaphorically, explaining the images in relative terms to readers who fail to grasp the more rigorous arguments. I set aside the choice of reading either literally or metaphorically and assert instead that the seemingly incompatible range of accounts of what-is in Fragment 8 should be understood instead in relation to kinds of structure. On my reading, Parmenides is going through the various levels of structure to provide the basic parameters for a complex account of causality for the sake of inaugurating a practice of natural philosophy appropriately oriented by the insight into being.[1] The result is a philosophy of nature oriented by, and appropriate to, the core insight into being in the Truth section, presented in a style that assiduously maintains the tension between the poem's primary insight and the apparently incompatible demands of causal explanation and natural philosophy.[2]

The Three Stages of the Reception of What-Is

Overview

The question of the relationship between the Truth and Doxa sections of the poem has been asked many times before.[3] I take as my point of departure the various markers that the poem has a circular composition, suggesting a departure from and return to human discursive understanding. In Fragment 1, the initiate departs from the human world in a chariot driven to its breaking point by "much discerning" mares guided by the daughters of the Sun, ultimately arriving at the Gates of Night and Day situated at the furthest limit of human knowing and presided over by "much-punishing Justice" (Δίκη πολύποινος, 1.14).[4] Once the divine guides persuade Justice to let them pass through those gates, the chariot moves beyond the structured differentiation between opposing kinds (Night and Day), and the initiate is welcomed and reassured by the goddess, as arriving at her side as is "just and right" (θέμις τε δίκη τε, 1.28). The goddess then announces that the initiate will learn about the heart of truth and about the "most estimable" (δοκίμως) opinions of mortals. In the Doxa section, the initiate has returned to the human sphere ready to discern and study natures with the appropriate orientation. The poem is thus divided into stages: initiatory passage, insight into being, and a return reoriented by that insight. The initiatory passage is captured in Fragment 1; the insight into being, which is commonly referred to as the "Truth" section, I frame as Fragments 2 through 6; the and the way of "Doxa" concludes the return (frs. 8.51 through 12). On my reading, Fragments 7 through 8.51 represent the transition from Truth to Doxa.

For reasons of length and thanks to the liberties afforded in the context of the present volume, I will presuppose the steps of interpreting Fragments 1 and 2 (cf. chapters herein by Mary Cunningham and Colin

Smith).[5] In brief, I am taking the ". . . is . . ." of Fragment 2 to have an elided subject and to call for completion by a predicate. Τό ἐόν ("what-is") is the noun form of the "is" of Fragment 2, which is completed as X, Y, Z, and so forth, that is, as the various true attributes gathered in Fragment 8. Thus, I take the ". . . is . . ." in Fragment 8 in the so-called speculative sense, as coined by Alexander Mourelatos, in which the attributes belong essentially to the subject (see Mourelatos 2008, 56–60). On my reading, being is the source of all truth in that it is what exceeds and calls for the determinately intelligible structure in things.[6]

I turn now to the arguments of Fragment 8 as framed by the goddess's Fragment 7 declaration of the "much-contested refutation," adapting a point made by James Lesher.[7] I read Fragment 8 as having the pedagogical purpose of reshaping mortal habits of thinking so as to preserve understanding as the initiate returns to the world of things. I take the refutation referred to in Fragment 7 to be the step-by-step reshaping of our understanding of the relationship between structuring cause and structured recipient. I will identify the various attributes of what-is as spelled out in the first few lines by the goddess, which I use as a provisional roadmap that is expanded in what I count as the three main sections of the fragment.

FRAGMENT 8, LINES 1–6

> Just one account of a route
> Remains: that . . . is . . . ; on this [route] there are signs,
> Very numerous, that what-is is ungenerated and
>   imperishable,
> Whole and of a single kind, and steadfast, and complete;
> it never was nor will be, since it is, now, all together,
> one, cohesive.
>
> [μόνος δ' ἔτι μῦθος ὁδοῖο
> λείπεται, ὡς ἔστιν· ταύτῃ δ' ἐπὶ σήματ' ἔασι
> πολλὰ μάλ', ὡς ἀγένητον ἐὸν καὶ ἀνώλεθρόν ἐστιν,
> οὖλον μουνογενές τε καὶ ἀτρεμὲς ἠδὲ τελεστόν,[8]
> οὐδέ ποτ' ἦν οὐδ' ἔσται, ἐπεὶ νῦν ἐστιν ὁμοῦ πᾶν,
> ἕν, συνεχές·]

Fragment 8 opens with a declaration that one account (μῦθος) of a route remains: ὡς ἔστιν (that . . . is . . .). The goddess says there are very many signs on this route that what-is has various attributes. By "signs," I take the goddess to be referring to images and arguments that show, that is, point to, these attributes.⁹

Each of the arguments in Fragment 8 works by denying the negation that belongs to one or another of the structures familiar to us in the world. The attributes fall into four basic categories, which are listed in 8.2–4: (1) in denial of the negation implied by a distinct precedent or consequent required by the paradigm of generation, what-is is ungenerated and indestructible (ἀγένητον and ἀνώλεθρόν); (2) in denial of the negation entailed by organized complexity, what-is is whole and of a single kind (οὖλον and μουνογενές);¹⁰ (3) in denial of the negation entailed by alteration, including developmental change, what-is is steadfast and complete (ἀτρεμές and τελεστόν); (4) in denial of the negation entailed by spatial and temporal structure, what-is is now all together (νῦν ἐστιν ὁμοῦ πᾶν), one (ἕν), and cohesive (συνεχές). These denials all in a sense say the same thing: the structuring cause cannot be subject to any of the characters that require nonbeing as part of their determinate intelligibility. The attributes themselves depend for their intelligibility on their opposites and what "is not," but in spite of being relationally determinate, these attributes are selected as appropriate based on an orientation toward and by the insight into being. At the same time, but conversely, each of these relational terms draws on and, thereby, highlights the basic terms to which our world is determinately intelligible.

At the first stage of reception (8.6–8.21), what-is shows itself as ungenerated because it, as the source of arising and perishing, cannot answer to any such source. That is a simple argument from a first cause, but it is specified in relation to phusis as the causal basis for determinate individuality and for the external relationships between individuals. At the second stage (8.22–8.33), what-is shows itself as a source of complexity, highlighted by the uniformity of distribution and the structuring limit depicted by the image of what-is "held fast in the bonds of" an enclosing limit. Thus, in addition to determinacy of the individual that arises and perishes, Parmenides now offers explanation for the internal structure of complex individuals. What-is here is fully established as the source of the specific nature of an individual, an object toward which thinking, naming, and production can secure their orienting terms.¹¹ The third stage of reception (8.34–8.51) is focused on the activities of naming and production, keying from the teleological orientation toward that "for the sake of which" (οὕνεκέν, 8.34) naming gathers itself and production orders its product. In this aspect, what-is is that "in reference to [which] all names have been spoken" (τῷ πάντ' ὄνομ' ἔσται, 8.38) and that toward which the production of the well-rounded ball must look for its model.[12] The introduction of the bulk of a well-rounded ball stresses the gap between the telos, the limit-imposed order, and that in which the order is set. Worldly order is oriented toward and by a telos, but that order is subject to variation and inconsistency (8.34–8.51). In conclusion of these three stages, the goddess declares the completion of the "trustworthy speech and thought concerning truth" (πιστὸν λόγον ἠδὲ νόημα ἀμφὶς ἀληθείης, 8.50–51) and goes on to describe the way of the Doxa (8.51–8.61).

The double character of these stages can be summed up as follows. In the first stage, we see that what-is is ungenerated, but at the same time we frame our causal account in terms of phusis. Similarly, in the second stage, we see that what-is must be undifferentiated in quality and quantity, but at the same time we are asked to imagine uniform spatial and temporal distributions on which our understanding of internal structure depends. In the third stage, the speech concerning what-is highlights the imperfections and deviations of the ordered in relation to its structuring source. Thus, while we are working to purify the insight into being, so as to orient ourselves toward what-is as the prior source of normative structure, we are also taking stock of ordered aspects of the world in which we live.

Notice that in all three stages, the attributes of what-is are gathered as if we were describing an object in the absence of any structuring cause, so an object absent any reason for a fixed ratio, division, or separation, in some cases simply a pristine receptivity without any cause to dictate earlier rather than later, no "right" time, no more rather than less, and so forth. This draws attention to the notion of a receptive continuum capable of being measured by the right amount or proper ratio, that is, a receptive basis for causally imposed structure. I am arguing that these are not only heuristic images helping us as signs to imagine what-is, though they might also be that. They are primarily elaborations of what is implied by the nature of a prior structuring cause, receptive to the right and proper time, ratio, place, and so forth. I argue that they serve as a positive contribution to the human reception of the goddess's wisdom in the ongoing study of natural phenomena.

Detailed Reading

I now turn to the textual examination of the three basic stages of reception of the goddess's insight, as outlined above. In Fragment 2, the goddess addressed the initiate and commanded the preservation of the insight into being: "having heard, *you* preserve the story" (κόμισαι δὲ σὺ μῦθον ἀκούσας, 2.1).[13] As she progresses through the stages of reception in Fragment 8, shifting between images and argument, the goddess provides the initiate with exercises through which to imagine and to think responsibly about what-is, helping us to understand what-is by transforming our established habits of thinking about being. In part, this is facilitated by the initiate being depicted as a character in the poem whom the reader recognizes as a surrogate, or whom the reader sees as looking for persuasion and as relying on authoritative conclusions, working through the elenchos.[14]

Reception of what-is by Justice (8.6–8.21)

As mentioned, Fragment 8's positive characterization of what-is through its attributes requires the negation of various types of non-being. The first set of types of nonbeing is drawn from phusis, specifically from the structures of birth and growth. At 8.6–7, the goddess asks, "For what parentage (γένναν) of it will you look for? / In what way, whence, did it grow?" She replies, "neither from what-is-not shall I allow / You to say or think [that what-is has either parentage or growth]; for it is not to be said or thought/That . . . is not. . . ." (8.7–9). Furthermore, what-is could neither "grow" (αὐξηθέν) from what-is-not (ἐκ μὴ ἐόντος) or "burst forth" (φῦν) from nothing (τοῦ μηδενός). She goes on to confirm that what-is must "be completely or not at all" (πάμπαν πελέναι . . . ἢ οὐκί, 8.11).[15] Along with other scholars, I take there to be two arguments here. In the first, at 8.7–9, the goddess denies growth from what-is-not, referring back to the conclusion of prior arguments,[16] and in the second, at 8.9–10, the goddess argues counterfactually that, had what-is burst-forth at some time from nothing, there is no reason, no rightful necessity, that would determine that it do so earlier rather than later.[17] The first argument draws on the insights of Fragments 2 and 6, and the second argument is a reductio that supports and reinforces the first.[18]

The ensuing section similarly combines two types of account, that is, the language of deductive inference and the imagistic portrayal of a goddess, Justice, at work considering evidence and ordering thought. The goddess first says that the "strength of proof" (πίστιος ἰσχύς) will not allow, from what-is-not, something to come-to-be alongside it (8.12–8.13).[19] She then shifts to overtly imagistic terms in lines 8.14–8.18, depicting Justice as placing what-is in "shackles," prohibiting what-is from coming-to-be and from perishing, "holding fast" what-is.[20]

Here I think we can appreciate the way Parmenides is emphasizing structure by drawing together argument and image, even to the point of eliding the difference between them. The goddess underlines that our inference follows from a decision (κρίσις, 8.15), just as she focuses on the act of having decided (κέκριται, 8.16). Justice constrains what-is and will not allow its coming-to-be or perishing (8.13–14); one "route" (ὁδός, 8.18) is allowed, so that (ὥστε) it "is" (πέλειν, 8.18) and "is true" (ἐτήτυμον εἶναι, 8.18). The other [route] (τὴν δ') is let go (ἐᾶν, 8.17) as unthinkable and unnamable. The reader, instead of simply engaging in the route of inquiry herself, is now following an anthropomorphized Justice acting as her surrogate. We readers are being asked to follow the decision made by Justice and to appreciate the dependence relationship ordering our inquiry (ἐν τῷδ' ἐστιν, 8.15, Coxon). The language of creed, the idea that the strength of proof or trust at 8.12 tends to bring the distinct register of deductive inference, especially distinct to us, closer to the figurative language of Justice making a decision, stressing their similarity. As distinct as inference and image are to us and to Parmenides, though surely to a lesser degree, Parmenides's decision to juxtapose deductive inference with image tends to draw the two closer together and emphasize the similarity between the structured nature of each.

I conclude from these observations that Parmenides is here facilitating the transition away from direct evidence by comparing various types of structured opposition, that of thought and image.[21] While the Fragment 8 project of gathering together the true attributes of what-is points back to the insight into being, the image of Justice making a decision to pursue one route and not the other points forward to the worldly domain of things and people.

At a higher level of interpretation, Parmenides is aiming to transform the habits of mortal understanding through the much-contested elenchos by drawing on a range of distinct but structurally comparable causal structures. The first stage of reception of the insight into being begins with phusis because this is a chief paradigm by which mortals understand causation, and the second stage repeats and reinforces the model of the agent-individual putting things in order. The much-contested refutation requires that we see being both ways, first by recalling the core of the Truth section and then drawing on the structures of mortal understanding. Overall, the elenchos is a reorienting return to the horizon of mortal opinion, one in which we learn to interpret various types of structure in a way that will allow us to understand how being can be causally responsible for things in the world.

As I mentioned above, some other scholars have explained the use of images in Fragment 8 as serving a pedagogical purpose of providing second-best argument to assist the initiate in turning away from mortal habits of thinking, the much-contested elenchos.[22] I contend that the arguments and images of Fragment 8 clarify the facets of the receptive basis for which what-is is causally responsible for ordering. The account of the true attributes of what-is is not radically distinct in kind from the causally governed birth of things, nor from the kairos for coming-to-be, nor from the movement to fullness and maturity. These are, one and all, ways of organizing mortal thinking in light of the insight into being.

Finally, let us consider the short coda to the first stage at 8.19–21, specifically for the way it educates us about the temporal structure of embodied life. The goddess denies that what-is could come-to-be in the future (ἔπειτα πέλοιτο, 8.19), could become (γένοιτο, 8.19), or could have come to be (ἔγεντ', 8.20), since the very nature of temporality is for each of these three (i.e., past, present, and future) to depend for their being on their nonbeing. These arguments clarify the nature of causality insofar as coming-to-be in time not only must be determined by some necessitation, some enjoining cause that calls for a specific moment rather than earlier or later, but it must also arise within a continuum of earlier-and-later, which is ordered to the kairos by a cause. Thus we see that what-is is not only prior to and different in kind from phusis and temporality, but that a cause governing birth and growth will also depend on temporal division between past, present, and future. I will continue this discussion in the following section.

In this section, I have argued that the initial stage of reception of the goddess's insight in Fragment 8 points ambiguously in two directions. When negating the nonbeing of the structures of phusis, the goddess is pointing to the necessity of ". . . is . . ." and the impossibility of "is not." This analysis resulted in the decision to allow what-is as the orienting insight along the route of inquiry. At the same time, through the images of phusis, the ordering hand of Justice, and the divisions of temporality, Fragment 8 elaborates the structure of the world and mortal thinking.

What-is constrained by Necessity (8.22–8.33)

In contrast with the focus on generation in the previous stage, the second stage of reception (8.22–8.33) focuses on the negations of divisibility, differential distribution into more-and-less or here-and-there, change, deviation, development, incompleteness, and, in short, any kind of self-opposition or internal complexity. These attributes, which all make sense given the priority of what-is relative to that for which it is responsible, follow directly from the conclusion reached in the previous stage. What-is "must be completely or not at all" (οὕτως ἢ πάμπαν πελέναι χρεών ἐστιν ἢ οὐκί, 8.11), which is repeated here as the principle that what-is must be all alike (πᾶν ἐστιν ὁμοῖον, 8.22), and is in turn confirmed by the principle that what-is "draws near" to what-is (πελάζει, 8.25).[23]

The various negations of nonbeing in the second stage are inferences from premises established in the first stage, which now serve as the origin from which the route of inquiry derives its course. The sudden lack of rhetorical questions, which are no longer being posed by the goddess, reflects the turn away from the core insight of the Truth section and toward an ordered path of inquiry as established in the previous section.

The goddess says what-is is "not divisible" (οὐδὲ διαιρετόν)[24] *since*, or *because* (ἐπεί), it "all alike is" (πᾶν ἐστιν ὁμοῖον, 8.22); it is "not somewhat more here [than there]" (οὐδέ τι τῇ μᾶλλον, 8.23), which would prevent it from "holding together" (συνέχεσθαι); "nor is it somewhat less" (οὐδέ τι χειρότερον), but "is all full of what-is" (πᾶν δ' ἔμπλεόν ἐστιν ἐόντος, 8.24).[25] From the premise that "what-is draws near to what-is" (ἐὸν γὰρ ἐόντι πελάζει, 8.25), the goddess draws the conclusion that what-is is "all cohesive" (τῷ ξυνεχὲς πᾶν ἐστιν, 8.25), which picks up the opening characterization of what-is as "all together" (ὁμοῦ πᾶν, 8.5). The goddess portrays what-is as "all full" (πᾶν δ' ἔμπλεόν, 8.24) and all cohesive, or as "changeless in the limits of great bonds" (ἀκίνητον μεγάλων ἐν πείρασι δεσμῶν, 8.26). Furthermore, the goddess says that what-is is "unbeginning and unceasing" (ἄναρχον ἄπαυστον, 8.27) "because" (ἐπεί) coming-to-be and perishing have been forced to wander far into exile, thrust out by "true trust" (πίστις ἀληθής, 8.28).

On its own, this last bit of reasoning seems to repeat the denial of generation from the first section, but we should notice that the goddess goes on to justify this claim on the basis of the "exile" or "straying" (τῆλε μάλ' ἐπλάγχθησαν/ἐπλάχθησαν, 8.28) of coming-to-be and perishing, which have been "turned back" (ἀπῶσε) by true trust. These images of turning back what-is-not could be taken as metaphors describing the insight that leads to the "decision" to allow what-is in the previous stage, but even as metaphors Parmenides is stressing secondhand thinking, that is, rule following. Thus, the necessity of the conclusions at this stage is hypothetical; it is as strong as the premises, which may either be seen apodictically or borrowed as a derived premise, or perhaps both.

In addition to hypothetical necessity and deductive inference, Parmenides expands the spatial imagery with the introduction of Necessity as an agent-individual imposing limits on what-is. Consider the goddess's argument that what-is "remains the same and in the same" (τωὐτόν τ' ἐν τωὐτῷ τε μένον) and "it lies in accord with itself" (καθ' ἑαυτό τε κεῖται, 8.29), thus remaining firmly in place (ἔμπεδον αὖθι μένει, 8.30), for (γάρ) Necessity holds it within enclosing bonds of a limit (πείρατος ἐν δεσμοῖσιν ἔχει, τό μιν ἀμφὶς ἐέργει, 8.31).[26] On this basis, she concludes, it is right (θέμις) for what-is to be "not incomplete" (οὐκ ἀτελεύτητον, 8.32), for it is not lacking, and if it were lacking, it would lack all. She moves without interruption between conceptual and spatial registers, as if to emphasize that thinking is here imagining itself oriented toward its premises. Conceptual structures are once again superimposed over spatial images, which I interpret to be emphasizing the commonality of the structure shared by images and arguments.

Finally, the goddess says that it is "not right" (οὐκ θέμις) for τό ἐόν to be imperfect, or it is right for τό ἐόν to be not incomplete, or apart from its end (οὐκ ἀτελεύτητον τό ἐὸν θέμις εἶναι, 8.32–8.33). If it were apart from its end or didn't have one, it would be lacking (ἐπιδεές). If it were [lacking], it would lack everything/all (μὴ ἐὸν δ' ἂν παντὸς ἐδεῖτο), repeating the insight that what-is cannot be subject to a prior cause that organizes partial phases of development.

To sum up, what-is continues to be thought and imagined in this stage as something subject to structures of opposition. There is no significant difference in kind between bodily uniformity or the cohesiveness, for example, of uniform circular or rotational motion, and the attributes of *being* selfsame and *not* beginning or ending. All these starting points lead to the same attributes. Whether we infer indivisibility from selfsameness or from uniform distribution and motion, we are in both cases thinking what-is in light of objects and in terms appropriate to mortal understanding, that is, the terms of relational structure. For these reasons, I have characterized the second stage of the reception of the insight into being as transitional to the Doxa section, in which worldly phenomena will be understood in light of structuring cause.

Before moving on, I would like to revisit my earlier claim that indefinite continua are not simply invoked as metaphors for thinking but are being formalized as fundamental parts of the causal structure that enables what-is to be responsible for the order of things. I take the explicitly indefinite nature of the continuum to be a conceptual achievement and something Parmenides is asking us to think through. Uniform distribution and rotational motion point to simplicity and perfection insofar as they express a kind of equanimity. The operative assumption is that differentiation would have to be explained, since it favors one over another, but nondifferentiation expresses an undifferentiated starting point, a kind of rightness or balance. The claims that what-is "remains the same and in the same" (8.29) and that what-is is "unbeginning and unceasing" describe a hypothetical object that is independent of any prior cause. The images of uniformity and equilibrium are images of what-is within the sphere of things and people, the sphere of opposites and determinate oppositions, ways of imagining what-is as something in and of time and space. When the goddess says that generation and destruction, which were extinguished and declared "unheard" in the previous section, are *exiled*, or have *wandered far away*, she is not simply enjoining our experience of spatial and temporal situatedness but purifying that content of causally imposed order. When she says of what-is that it is "no more here" nor "less there" but "all full" and "cohesive" and "unbeginning" and enclosed "all around," she is exercising our understanding of the unformed and unstructured, which is appropriate to what-is, which has no prior cause governing it. Why do this? Because the more precise we are about the attributes of the what-is, so to speak, the more explicit we become about the reception of structuring cause.

In conclusion, I have argued that the core Parmenidean insight is that being is prior to structure, but our reception and preservation of this insight resides within mortal oppositions, specifically within the way we conceive of casual structure. Thus, the goddess moves freely between structures of predicative thought, inferential reasoning, shape, and motion. At the level of divine insight and at the level of mortal reasoning, Parmenides clarifies the nature of the needful, but in different degrees. He starts with the strongest evidence that ". . . is . . ." must be and "is not" cannot be brought about, and then he moves to hypothetical necessity and the entailments that can be expressed as, and follow from, an image. I have traced the goddess's shift from the rhetorical questions of the first stage to the ordered inferences of the second stage. The sense of the needful (χρεών), has moved away from direct engagement with evidence to what follows appropriately as fitting entailment. In the next section, Parmenides extends the account into the explicitly mortal context of language and contested meanings.[27]

What-is subject to Fate (8.35–8.51)

The third stage of reception (8.35–51) describes the divine insight as it appears in the mortal realm subject to Fate (μοῖρ', 8.37). The goddess compares what-is to a well-rounded ball (εὐκύκλου σφαίρης), that is, as an object of technē, and she stresses mortal naming. As I read it, the concern in the previous two stages was to outline what-is as a theoretical object, to the degree that it is appropriate and possible to do so, by reasoning out its attributes and sketching its image. At this stage, the concern is to situate that insight in relation to imperfection and instability of mortal productive activity.

As I have argued, truth in the strictest sense is the insight into being, and the true attributes of what-is are secondary. Whether we are organizing true attributes in image or concept, we are in both cases working up a structured articulation of an insight that is prior to structure. In the previous section, Parmenides imaged what-is in spatial and temporal terms as perfectly self-contained, at rest within itself, and uniformly distributed. The shift in the third stage to imperfection and instability is flagged by a turn toward a likeness of what-is. A "likeness" is a relation that expresses the difference between individuals possessing the same attributes, including the difference between a thing and the model on which it is based. Parmenides uses the image of a "well-rounded ball" as a likeness of what-is. The bulk of the sphere participates in perfect attributes but also admits of imperfection or imbalance. The well-rounded ball serves as an image of something that is ordered in relation to the aim or telos of the craft, but which can fall short due to the variation of which a bulk naturally admits. This contrasts the discussion of structuring limits in the previous sections, which did not foreground the imbalances into which an ordered individual can fall.

Turning to the text, Parmenides says that what-is is bound by Fate (μοῖρα), and he uses the image of a productive technē to elaborate the ordered activities of persuasion and naming:[28]

> . . . for nothing else is or will be
> Besides what-is, since Fate has bound it
> To be whole and changeless. Therefore, with reference to it
>   all things have been named,
> All these that mortals, believing in their truth, have
>   established;

'To come-into-being' and 'to perish,' 'to be' and 'not to be.'
And 'to change place' and 'to transform in bright color.'

[... οὐδὲν γὰρ <ἢ> ἐστὶν ἢ ἔσται²⁹
ἄλλο πάρεξ τοῦ ἐόντος, ἐπεὶ τό γε μοῖρ' ἐπέδησεν³⁰
οὖλον ἀκίνητόν τ' ἔμεναι · τῷ πάντ' ὀνόμασται
ὅσσα βροτοὶ κατέθεντο, πεποιθότες εἶναι ἀληθῆ,
γίγνεσθαί τε καὶ ὄλλυσθαι, εἶναί τε καὶ οὐκί,
καὶ τόπον ἀλλάσσειν διά τε χρόα φανὸν ἀμείβειν.]
(8.36–8.41)

What was previously bound by Necessity, here bound by Fate, represents what-is ordered in relation to a telos. Fate "has bound" (ἐπέδησεν) what-is to be whole and changeless (8.37–8.38), determining what is mete and right. Just as Fate implicitly looks toward something when ordering what-is, so too we look toward what-is when establishing the names that organize our world. Mortal opinion, which establishes itself in and through the act of naming, is not only two-stages removed from direct insight and decision, but is associated through the image of the well-rounded ball with the potential for imperfection and errancy.

The lengthy description of what-is on the model of the bulk of a well-rounded ball at 8.42–8.51 brings to the fore the relationship between structuring limit and the receptive base on which limit and measure are imposed:

> Since, then, there is a furthest limit, it [i.e., what-is] is perfected
> <From all sides> like the bulk of a well-rounded ball,
> Equally balanced <from the middle> in every direction; for it must not be any larger
> Or any smaller here or there;
> For neither is there what-is-not, which could stop it from reaching
> Its like; nor is there a way in which what-is could be
> More here and less there, since it is all inviolate;
> For equal to itself in/from every direction, it lies uniformly within limits.

[αὐτὰρ ἐπεὶ πεῖρας πύματον, τετελεσμένον ἐστὶ
πάντοθεν, εὐκύκλου σφαίρης ἐναλίγκιον ὄγκῳ,

μεσσόθεν ἰσοπαλὲς πάντῃ · τὸ γὰρ οὔτε τι μεῖζον
οὔτε τι βαιότερον πελέναι χρεών ἐστι τῇ ἢ τῇ.
οὔτε γὰρ οὐκ ἐόν ἐστι, τό κεν παύοι μιν
   ἱκνεῖσθαι
εἰς ὁμόν, οὔτ' ἐόν ἐστιν ὅπως εἴη κεν ἐόντος
τῇ μᾶλλον τῇ δ' ἧσσον, ἐπεὶ πᾶν ἐστιν ἄσυλον·
οἷ γὰρ πάντοθεν ἶσον ὁμῶς ἐν πείρασι κύρει.]

The bulk of the ball is described as not being larger or smaller "here or there" (τῇ ἢ τῇ), which invites us to think of balance and isometry of a bare spherical continuum in which there is no cause ordering a ratio of more rather than less, here rather than there, and so forth.[31] In recent years, some scholars have taken the comparison to the bulk figuratively as a reference to the process of working an entire circumference into equal tension all around, so that no part of it bulges out lopsidedly (cf. Mourelatos 2018, 136).[32] The image fits well with the goddess's previous efforts to provide images of a continuum in equipoise, lacking governing cause, and, thereby, lacking definite ratio. As described, the ball is characterized by equipoise "from all sides" (πάντοθεν) and is an aperture looking onto the idea of what-is as perfected (τετελεσμένου), equally balanced (ἰσοπαλές), with no lopsided bulges to make it unlike itself; it is "inviolate" (ἄσυλον), which I take to mean that it is symmetrical and balanced. Thus, this image fits the general purpose of the goddess, previously evidenced, to encourage us to think of an object in the absence of any governing cause. In my view, however, this image also brings the nature capable of spatial uniformity to the fore.

My interpretation departs from most others in stressing the goddess's introduction of the bulk (ὄγκῳ) as a way to thematize receptivity to ordering limit. A bulk is something that can become well-rounded, but of course it can also fail to be, and it can harbor imperfections. Rather than being absorbed into and exhausted by previous images of spatial and temporal continua, the image here transforms those earlier examples and brings out what I see as their secondary purpose, which is to require of us readers that we formulate for ourselves the notion of a continuum capable of being ordered toward the right amount or right ratio, something that has no inherent proportion but that works with what-is to make ordered proportion possible. We began, as I argued above, with the spontaneity of being, and we conclude with an image drawn from technē that allows us to imagine being shaped by what-is as its cause.

The Parmenidean Elenchos (8.51–61)

At 8.51, the goddess begins speaking explicitly in terms appropriate to mortal thinking, picking up where she left off at the end of the Proem. She warns us that her words will be deceptively ordered. Mortal thinking, the goddess says, relies on opposing and apparently separate terms. For instance, and in general, the qualities of being light and selfsame seem to be separate from the qualities of being dark and heavy (8.56–59). When the goddess generalizes these opposites, I take Parmenides to be describing what-is as the unity of forces of combination and dispersal, illumination and obscurity, that structure our world in light of being.[33]

Looking back to the above observation that the elenchos referenced in Fragment 7 points forward to Fragment 8 and its various arguments, especially to the decision Justice makes in Fragment 8, we can expand on what it means to demand that we see and judge this elenchos for ourselves. That exercise has consisted of differentiating what-is in stages, beginning with seeing being on its own terms in the Truth section of the poem, and then in Fragment 8 specifying what-is as a source that gathers its attributes, continuing from there to appreciate the demands of causally structuring relationship between what-is and the order of things in the world, and finally ending with the indefinite continua or more-and-less on which causes impose structure but that express their own indifference to structure.

I have presented an interpretation of what-is as passing from the insight into ἔστι that is originally evident in prior fragments into three stages of reception and preservation in Fragment 8: (1) structured as what-is, being orients a route of inquiry; Justice has decided to allow the true route, which invites us to adopt a thoughtful surrogate for prior insight from which we are now deriving inferences and ready to study the structure of generation; (2) oriented by cause, which is itself bound by strict necessity, we study and appreciate the complexity imposed and solicited by structuring cause; and (3) as the telos that guides the ordering of names and products, what-is is imposed on an indefinitely receptive natural basis. I have organized the various images of equipoise and uniformity as invoking an indefinite continuum that becomes gradually more explicit as we approach the goddess's image of a bulk or mass (ὄγκος). I have focused especially on the way the goddess characterizes what-is as a cause, using images to draw to mind the notion of an indefinite continua of more-and-less.

To review and clarify: Parmenides has used the image of indefinite, uniform continua in three cases: (1) in the second part of the long argument against generation (8.5–21), the goddess uses the idea of a continuum of earlier and later to argue that there is no reason why what-is, had it arisen from nothing, should arise at one time rather than another; (2) in the argument concerned with being whole and of a single kind, unbeginning and unceasing, the same and in the same (8.22–33), she argues that what-is is indivisible because it would rightly be "no more here" (οὐδέ τι τῇ μᾶλλον) "nor any less" (οὐδέ τι χειρότερον);[34] and, finally, (3) in the closing section concerning completeness and perfectedness (τετελεσμένον, 8.34–8.51), she compares what-is (ἐναλίγκιον) to the bulk of a ball that has been well-rounded and properly balanced, as we have just considered.[35] These images of indefinite continua are combined in the description of what-is as shackled (8.14), situated within limits of great bonds (8.26), enclosed from all around (8.31), bound (8.37), having a furthest limit (8.42), and ultimately as evenly balanced from every direction (8.44), within limits (8.49). I have argued that these indefinite continua of more-and-less are not merely heuristically helpful images of what-is but are an independently necessary part of the Parmenidean conception of the causally structured cosmos, specifically the space, time, or bulk, that are receptive of measure and capable of right ratio, right time, right coming-to-be and perishing, right naming, and so on. By focusing on these images, we are now in a position to understand them as preserving our insight into what-is through a principled study of nature attuned to structuring cause.

Part of the evidence for my reading is simply the number and precision of Parmenides's spatial metaphors. There is insufficient motivation to use such metaphors so frequently unless they were an important part of the elenchos, and there would be even less motivation to progress step-by-step to more explicitly spatial metaphors unless spatiality (and temporality) were themselves to be stressed. As we read, and as various aspects of what-is are being imaged, we are being initiated into thinking about the absence of cause, both in the sense of what it would mean to "be" without being in determinate relation to an other and in the sense of that upon which what-is imposes itself. The goddess's references to indefinite continua of time, space, and spatial distribution/mass continua provide a key conceptual component of knowledge of the world, a knowledge that we access through the goddess's much-contested elenchos distilling aspects of our ordinary, habitual ways of seeing and relying on the world.

As remarked above, what-is, understood as the causal source of order in the world of things, is determined according to a series of denials of negation. Once we remember that Justice also has the job of deciding who passes through the Gates of Night and Day into the divine realm "as is Themis and Dike" (θέμις τε δίκη τε, 1.28), we put ourselves in a position to appreciate the degree to which the terms of our thinking remain tied to objects. Much-punishing Justice (1.14) holds apart opposites in the mortal realm and guarantees their structured interrelation. Justice in Fragment 8 shackles what-is and establishes the route of inquiry along which what-is, as a particular instance within the horizon of opposites, can become present to thinking as something that points beyond that horizon, as a cause in which to place trust (πίστις) and as a basis for persuasion and estimable judgment.

This latter image carries unmistakable echoes of Justice's role as polupoinos: Justice makes room for one opposite (i.e., what-is) by forcing away the other (i.e., the interrelation of opposites). To wit, the process of "bringing near" what-is has allowed us to imagine what it would mean, and how it would look, to be devoid of prior cause, some "thing" that cannot be subject to partiality, temporal and spatial differentiation, movement, change, development, opposition, being scattered and combined, and most generally, coming-to-be and destruction. Then, by elevating indefinite magnitudes of various kinds—hot/cold, wet/dry, more/less, before/after, here/there—to a level of explicitude as that upon which the limits are imposed by τό ἐόν, the Doxa section is able to orient our thinking, perceiving, and speaking toward the cause that determines the proper time for generation, the proper distribution or delimitation of more-and-less along various continua, or the articulation of structure. As Parmenides indicates, the purpose of the Doxa section is to establish "a likely world-scheme" (διάκοσμον ἐοικότα, 8.60) so that the initiate will have judgment that cannot be outstripped by other mortals. The process of carving out a space for cause and continuum is the basis on which we become capable of achieving that account of "likely order."[36]

In response to the challenge that, as others have interpreted Parmenides, the goddess is either speaking merely metaphorically wholly literally,[37] I would agree that Parmenides engages with what we call metaphor to re-route mortal habits, which takes time and patience. To the literalist, I respond that Parmenides's survey of causal structures conceives the uncaused in a variety of distinct ways, including as an indefinite continua in bulk/extent, time, and place, to challenge us to think through the receptivity demanded by structuring cause. Fragment 8 exercises mortal opinion by articulating the stages of determinate intelligibility, down to limits of place, time, position, the scattering and solidifying, or concentration and dispersal, and so forth.

Conclusion

I have in this chapter elaborated an account of what-is (τό ἐόν) according to which the various images and arguments in Fragment 8 can be understood as stages in the reception of the insight into ἔστι of Fragment 2. By focusing on a shift from the present active form of the verb "to be" (ἔστι) to the same verb in noun form (τό ἐόν), I have interpreted distinct ways in which the insight into being is preserved in logos once it has been constituted as a theoretical object along a route of inquiry, taken up as present to thinking, and thereby capable of functioning as the normative structure guiding naming and explanatory judgment. In each one of these preserved modes, ἔστι is present as rightful (due, appropriate) necessity. At the same time, the images of uniform distribution in space, time, bulk, and so forth, have been shown to have the dual function of pointing toward what-is and calling for newly focused understanding of indefinite continua as that on which causal order is imposed. I have argued that the process of becoming explicit about what-is involves us in an exercise of rehabituating our relationship to things. My reading of Parmenides's method explains the extensive and progressively explicit use of spatial and temporal continua as an important part of that rehabituation.

I have argued that by appreciating the stage-wise modulation of the reception of the insight into being, Fragment 8 allows us to reorient our study of nature in light of the poem's core insight. Posed with the question whether being is singular or plural, and whether the ". . . is . . ." may be the prior source of a whole noetic/causal context or just that context itself, I have charted a middle course, remaining agnostic regarding whether what-is consists of a plurality of determinate causal sources into which inquiry can probe. In my view, Parmenides, unlike Plato, does not subject being, loosely speaking, to the limits and relationships necessary to understand the combinations and divisions between ideas. Instead, what-is is conceived as a single source from which any number of hypostasized causes might be derived, maintaining the distance between being and things. The priority of being to the number and determinate nature of kinds, for instance, suggests that Parmenides's focus is on preserving the insight into being as an orienting basis on which to conduct the study of nature. He is not involved in a project of mapping out the objects of contemplation in a theoretical or theological space. In virtue of seeing the world of things as subject to limit, and as rooted in indefinite continua (e.g., of more-and-less, here-and-there, earlier-and-later), we can appreciate the source of fittingness, that is, the source of the kairos, in things. Thus, I conclude that our understanding of the principles responsible for the conditions of things in the world might aid our causal explanation of structure in the world, but Parmenides's goal is not technical mastery of worldly phenomena, nor is it pious preservation of a fixed and given order. His goal is to expand our appreciation of the profound depth of being at work in the many-layered surfaces of causally imposed structure.

Notes

1. The evidence that Parmenides is responsible for a theory of heliophotism (i.e., that the moon is illuminated by the light of the sun), for the identity of the morning and evening star, and related basics of astronomy are some parts of the evidence of Parmenides's sincere interest in natural philosophy. This has been well remarked by Cerri (2011), Curd (2004), Cherubin (2017), Graham (2006, 2013), Gregory (2014), Mourelatos (2008, 2012, 2020), and Rossetti (2020). My view is that his interest is fundamentally oriented by the desire to appreciate the presence of the divine as expressed in the normative structuring of indefinite continua. The Doxa section should be understood not in light of the predictive certainty or the reliable manipulation of the world yielded by knowledge and art (τέχνη) in natural philosophy but, instead, in terms of the mortal appreciation of the depth of the real behind the surface of things. Such practice of inquiry would establish the most trustworthy and estimable of all mortal knowledge.

2. I remain agnostic with regard to the potential plurality of determinate causal sources. My view is that the study of being is a study of causation, and Parmenides is clarifying the nature of being as cause. However, Parmenides is strict about the radical difference in kind between being and all structure, whether qualitative or quantitative, because all of it entails nonbeing and is situated within the horizon of mortal thinking. As has been established, there other ancient Greek philosophers who proposed an array of different causal regimes without, apparently, believing themselves to have abandoned the Parmenidean insight. I think these other philosophers are responding to Fragment 8 and the way of Doxa generally, and that, strictly speaking, Parmenidean "being" refers to the normativity itself that is presupposed in all structure, especially spontaneous and essential structure.

To indicate that being as the normative source of structure is not itself subject to structure, I will refer to it as "the structuring cause." My approach, which is to take being as a cause, is consistent with a suggestion made recently by Patricia Curd: "I am beginning to wonder if what-is for Parmenides might turn out to be less an entity (thing) and more a principle or set of principles that describe and order the cosmos . . . Parmenides may suggest that the permanent and unchanging being of the principles that govern all things provides that explanation [of, e.g., the way the sun and moon move in predictable ways]" (Curd 2015, ¶ 31). Curd takes this construal to shift Parmenides in the direction of Heraclitus's account of the logos as the set of laws of the cosmos.

3. Against my view that the two parts of the poem are continuous, see Rossetti's critique (2020) of attempts to connect the two and for the view that the two parts of the poem are not substantively continuous. In this volume, Jeremy DeLong, Jessica Elbert Decker, and Sosseh Assaturian take approaches similar to mine.

4. It is an editor's decision whether to capitalize δίκη, choosing to refer to the goddess, or to leave it lowercase, referring to the abstract noun. For example, Coxon leaves it lowercase and here translates "retributive justice" (2009, 50–51), and Mourelatos capitalizes Δίκη and translates "much-punishing Justice" who holds the keys "of retribution" (ἀμοιβούς) and relates πολύποινος, as "much-avenging," to its Homeric context (2008, 5, 15). Some scholars point to five specifically female deities in the poem, Δίκη, Ἀνάγκη, Μοῖρα, Θέμις, and the unnamed truth-speaking goddess (θεά) who receives the initiate; that total becomes six if we include the daughters of the Sun. On my reading, only two goddesses, Δίκη (who decides to allow the route of inquiry) and the unnamed θεά (who questions and converses with the initiate), exercise agency and thereby should be considered goddesses.

5. Following and adapting work by Charles Kahn (2003), Lesley Brown (1994), Alexander Mourelatos (2008), and Mitchell Miller (2006), I am taking the verb "to be" to have a tacitly locative force as being-there-for-thinking, which I intend to show is compatible with the structure of things in the world, bridging the apparent gap between being and nature. We conceive being "directly" or "primarily" when, for instance, the goddess invites us to try to bring about the opposite of being, pointing out that unqualified nonbeing (τό γε μὴ ἐόν) cannot be brought about (ἀνυστόν, 2.7). Reading the γε in τό γε μὴ ἐόν as an intensifier pointing to nonbeing "as such" is not a common reading of the Greek (cf. Miller 2006, 3–4). We can see this for ourselves. By contrast, we preserve the ἔστι directly on the basis of the impossibility of bringing about nonbeing as an object of thought and inquiry, that is, as manifestly incapable of apprehension by mind. Being is what brings about, and it can itself be "brought about" in the sense of being apprehended as the power to compel thought and guide logos. However, once we establish this power of being through its attributes—which we must as the ἀρχή of further inquiry—the very notion of an articulating contrast between

being and nonbeing, subjected to the structuring demands of logos, misses the radical necessity of ἔστι. For accounts of nonbeing as nondestination of inquiry or as a lack of guiding power, cf. Mourelatos 2008, 328; and Curd 2015, ¶ 10. Note that I am not emphasizing the experience of thinking as opposed to the "object" of apprehension, but I appreciate the appeal to direct evidence of the impossibility nonbeing made by Robbiano (2016).

    6. I note that Heidegger construes νοῦς and νοεῖν in terms of apprehension by mind (vernehmen, 1976 [1953], 105), and takes "being" as what gathers and organizes predication as a source of fitting together; from the verb fügen, to put together, to join, he writes "λόγος, ist fügender Fug: δίκη" ("logos is [a] fitting joint: justice," 1976 [1953], 23); additionally, Logos is "das Verhältnis des einen zum anderen" ("a relationship, or ratio, of one to another," 1976 [1953]: 95). For similar accounts of what-is as the basis for thinking, structured by logos, but nonetheless not identical to logos, cf. Cordero 2004, 162 and Mourelatos 2008, 326–30. For an account of ἔστι in these passages that is strictly focused on the sense of the copula in necessary statements of identity, see Coxon 2009, introduction, section 5.

    7. With Lesher (1984 and 2002), I interpret Parmenides's Fragment 8 with a focus on the πολύδηρις ἔλεγχος referenced in Fragment 7, though I take Fragment 8 in a way he does not. On my view, our attempt to single out what-is as structuring cause is a matter of preserving the insight into being in a way that can reorient the study of nature. Mourelatos (2020, 249–50) rightly, in my view, has taken Parmenides's discovery of being as a passage beyond images into a register free from "anthropocentric familiarities." Unlike Mourelatos, who accords the images of Fragment 8 a limited status as an "indirect proof" (i.e., reductio) employing a sort of οὐ μᾶλλον principle, that is, "no more" this than that, which he has associated with the Principle of Sufficient Reason (Mourelatos 2008, 160), I take Parmenides's use of indefinite continua to be metaphysically pointed, and so not strictly metaphoric and heuristic. I borrow Mourelatos's term "modulation" to refer to different degrees of evidence with which the goddess presents what-is, from direct insight to persuasive images (Mourelatos 2008, 84; cf. also Mourelatos 2018, 130–31). Whereas I follow Mourelatos's stress on the progressively explicit use of spatial imagery (Mourelatos 2008, 128–29; noted by McKirahan 2008, 214 and 228 n. 91), I read the spatial imagery as both a pedagogical tool and as a basis for a proposed study of causal structure.

    8. Coxon 2009, 315: "The best emendation is ἠδὲ τέλειον ['and perfect'] (G. E. L. Owen))." However, Palmer (2009, 383), referring to 8.32 and 8.42–3, removes the motivation for emending the text.

    9. McKirahan (2008, 221 n. 9) argues that the "signs" are not themselves the attributes but rather the arguments indicating what-is to have its attributes. By "argument," he is referring to the logical connections drawn throughout by the logically connective terms, including "for" (γάρ), "since" (ἐπεί), "therefore"

(οὕτως; τῷ; also τοῦ εἵνεκεν, γάρ), "thus" (οὕτως), and "because" (οὕνεκεν). For the logical structure signaled by the particles of Fragment 8, see Mourelatos 2008, 174. The other obvious way to take "signs" would be as the various attributes assigned to what-is, each attribute being a sign of what-is; the function of the signs would thus be to allow us to pick out what-is. Curd (2004, 48 n. 68), e.g., writes that the "signs" refer to criteria by which we pick out what-is, which she construes as the natures sought in the context of Ionian philosophical inquiry. For an account of the truth of what-is as primarily a logical, rather than ontological, inquiry, see Wedin 2014. I refer to "images and arguments" because it is central to my reading that Parmenides is effacing the distinction between image and argument to emphasize structure.

10. Some others translate μουνογενές as "of a single kind" (cf. Mourelatos 2008). McKirahan argues in favor of "unique" (McKirahan 2008, 221 n. 10).

11. Because Parmenides purifies complex structures of the being immanent to them, he is able to refer to being as if it were a mass noun, both divisible and combinable as something, for example, that "draws near" to itself, but also as undifferentiated as "simple" and "unique."

12. There is a textual dispute here, but either of the main senses from which we must choose would give the same sense of mortal names owing themselves to what-is.

13. There are objections one might raise to the account itself as a violation of the prohibition on what-is-not. One might object that to define what-is in relation to what it is not entails, since negation is symmetrical, either that it is not [its other] or that its attributes are not [their opposites]. This would be to treat the logical relations between attributes as if they were subject to the oppositions of "much-punishing Justice," which the initiate left behind when passing over to divine insight. I think this is not misguided. Fragment 8 remains beholden to the domain of logical opposition insofar as we treat ". . . is . . ." in Fragment 8 as subject to logos and name it "what-is." Logos inescapably pairs opposites as the same in kind, which is what will prompt the goddess's various images of "limit" as a corrective (e.g., "shackles" [8.14], "holding fast" [8.15], a spatialized sense of κρίσις [8.15–8.18], "great bonds" [8.26], "exile" of "becoming" [8.27–28], "in place" [8.30], "bonds of a limit" [8.31], "fate has bound it" [8.37], "furthest limit" [8.42], "well-rounded sphere" [8.43], and "uniformly within limits" [8.43]).

14. Cf. Cunningham's chapter in this volume on initiation in the poem.

15. She introduces this thought with οὕτως, which might in other contexts be taken to be drawing a conclusion; in this context it seems to function more as a confirmation of the premise that nonbeing cannot be. For an interpretation of Fragment 8 centered largely around the notion of "being completely" or "being fully," see McKirahan 2008.

16. I follow McKirahan (2008, 223 n. 20), who takes this "from what" to refer to what turned into it, as a seed turns into a tree. McKirahan takes "birth" and

"growth" to be the same, and Mourelatos takes birth and growth to be different. I take "what parentage?" to refer to both arguments, with "growth" referring to a transformation from a prior and distinct stage of development and "burst into being" referring to the appearance of a being wholly new (Mourelatos 2008, 244; cf. also esp. Mourelatos 2018).

17. Adopting the translation of Pulpito and Mourelatos (2018). There is an extensive discussion concerning the variety of translations of φῦν and χρέος (cf. Mourelatos 2018). ὦρσεν (from ὄρνυμι) can have the sense of an internal compulsion or need, and it is the root of our word "hormones." However, more commonly it can also have the sense of calling forth; the LSJ cites examples in Homer of storms being called forth by gods. The implication is that there is a context for such a call, and it has a place. Mourelatos (2018, 121) suggests that we read χρέος as a nominalization of χρέ and χρεών ἔστι, conveying "right and reasonable necessity."

18. Pulpito and Mourelatos (2018) read various aspects of Fragment 8 as either metaphorical or a reinforcing argument, strategically implemented to strengthen the goddess's pedagogical project with arguments of different kinds. They write, "Even though the rejection of what-is-not is the overriding argument throughout B8, and is explicitly cited at B8.7–9 and again at B8.46, the goddess chooses to reinforce it in both contexts (B8.6b–10 and B8.44b–B8.48) by the quite different argument that invokes [the Principle of Sufficient Reason]" (Pulpito and Mourelatos 2018, 134–38). I think these references to indefinite continua are essential to the project of differentiating what-is from the conventional understanding of beings.

19. McKirahan takes αὐτό here to refer to what-is-not but acknowledges that we might also want to take it to refer to what-is (McKirahan 2008, 223 n. 22; cf. Mourelatos 2008; Curd 2004 [1998], 78, cited in McKirahan). Whether we take "it" here to refer to what-is-not or what-is, either of which is grammatically possible, the lesson is the same: there can be neither origination nor growth of what-is because what-is cannot be subject to difference and opposition.

20. "Right and justice" (θέμις τε δίκη τε) and "no ill fate" (οὔ τί . . . μοῖρα κακή) sent the initiate through the gates of night and day to the presence of the goddess in Fragment 1. They reappear in Fragment 8 as agents thinking and acting in accord with necessity (ὥσπερ ἀνάγκη).

21. Even Parmenides's language used here to denote "being" makes tacit reference to spatial structures. When Justice makes a decision to allow what-is, so that it is and is genuine, the "is" in that result clause translates πέλειν, which can indeed be taken as a synonym for εἶναι when supplemented by a predicate adjective, as I have here, but that on its own also means "come to be," and is related to πελάζειν ("to come near," cf. 8.25). She concludes: "Thus coming-to-be is extinguished and perishing not to be learned of" (8.21). The very language implies relationships of opposition between near and far, familiar and strange, in terms of which the true result of the movement of thinking is cast as something

produced and preserved. Furthermore, notice the consistent use of practical imagery to identify what is not "allowed" (ἐφίημι, 8.12) by trust, what is not "allowed" (ἐάω, 8.7) by the goddess, and what is "allowed" (ἀνίημι, 8.14) by Justice. With the possible exception of "strength of trust" (πίστιος ἰσχύς, 8.12), the movement of thought is depicted in third-person terms in which the image of the agent-individual dramatizes what we would commonly think of as the movement of thinking. In the case of "trust," we might take "allow" to refer to the givenness of what-is to thinking.

22. The verb "decide" (κρῖναι) is a second-person, middle-voiced imperative, carrying the sense of "judge for yourself"; though compare Jenny Bryan's chapter in this volume.

23. Though we might be tempted to hear Themis personified in a way reminiscent of the poetic tradition, especially given the agency of Necessity in this stage and Justice in the previous, the term is here used as a predicate adjective and not, I would say, as an agent either deciding or shaping what-is.

24. As noted by McKirahan (2008, 223 n. 31), "διαιρετός means both 'divisible' and 'divided,'" which he understands in light of commitments stemming from his reading of the later claim "what-is draws near to what-is" (8.25).

25. McKirahan (2008) takes a different approach, translating τῇ as "in any respect" and translating πᾶν ("all") and ὁμοῖον ("alike") as adverbs qualifying how what-is has being. That reading makes clear sense of the argument. However, consider that when Parmenides says what-is is "fully," he says that if it lacked at all, it would lack everything. It would lack everything because, rather than being what-is, it would be subject to what-is, that is, it would be radically distinct from itself. To say that it is "no more here" "nor any less" is the same as saying what-is is "now," between before and after. McKirahan writes, "I take πᾶν ("all") and ὁμοῖον ("alike") in the second half of 8.22 to be adverbs," and, "Here the words are unquestionably adverbs."

26. David Sedley argues that taking Parmenides's references to motion in a nonliteral sense faces the danger of "diluting the argument into the trivial," and that Parmenides's references to "larger" and "smaller" at 8.44–8.49 function as parts of a real argument if taken in "a literal spatial sense" and should be taken literally in the absence of a plausible metaphorical interpretation (Sedley 1999, 119, 121). I do not feel confident that the literal/metaphorical distinction as we understand it applies here without qualification. The truth is stranger. I agree that our interpretation should take the notion of various spatial and temporal continua seriously and as more than a metaphor pointing toward what-is, but I think that Parmenides groups the predicative structure of attributes together with what Sedley considers the "literal" account of spatial and temporal attributes, all of which we should think of as expressive of mortal thinking in terms of opposites.

27. Mourelatos (2018, 133) distinguishes the other modes of to χρεών from Necessity as follows: "the other four have the more genial face of what is 'right,' 'proper,' 'appropriate,' and 'fitting': Themis te Dikē te (B1.28), Dikē (B8.14), Moira

(B8.37), and, most importantly, Peithō, 'attractive cogency, persuasion' (B2.4; cf. pistis alēthēs at B1.30, B8.28)."

28. The image of "fate" (μοῖρα) should perhaps be personified as the goddess who bestows fate, lot, destiny. If so, then there are a range of senses in which this might be taken. I accept Coxon's suggestion that we have here a reference to Homer, who uses the expression μοῖρα πέδησε three times, specifically referring to *Iliad* 10.5, where that phrase is followed, as it is here, with an epexegetic infinitive (Coxon 2009, 11). The meaning of the reference to fate here, as I understand it, is to the fitting or appropriate, mete and right, as in Homer's sense of something that is κατὰ μοῖραν. I take this usage to parallel Hesiod's account of the Moirai as children of Zeus and Themis at *Theogony* 904. This usage is distinct from the Fragment 1 reference to μοῖρα κακή, which would seem to be a reference to the road that one follows at death.

29. Adopting McKirahan's text, following DK (McKirahan 2008, 225 n. 48).

30. McKirahan 2008, 226 n. 63.

31. McKirahan translates τῇ ἢ τῇ as "than in another," and Coxon as "in one regard than in another." These have very different senses, but both allow comparison of what-is to what-is, urging us to see sameness in uniformity.

32. Owen and Guthrie represent opposing traditional views. Both think of Parmenides as attempting to carve out a new insight using available terminology. Whereas Guthrie considers the relationship between circular motion and "psychic function" to be an "analogy" (Guthrie 1962, 43), he considers the shape of the sphere to be, more than an example, a description of what has no outside (Guthrie 1965, 43–49). Alternatively, Owen takes Parmenides to be in the bind of needing to use spatial and temporal language to rule out spatial and temporal variation (Owen 1960, 100). For literal readings, see Gregory 2014 and Sedley 1999, 113–33.

33. For a lucid discussion of the distinction between Light and Night as it incorporates density/rarefaction and applies as determinate contrast to all perceptual experience and spatio-temporal determinacy, see Cherubin 2005, 1–23. For Parmenides's earlier reference to combination and dispersal, cf. Fragment 4: "Look at things though absent for the mind present securely:/for you will not cut off what-is from clinging to what-is,/since it is neither completely scattered (σκιδνάμενον) everywhere in the world/nor combined (συνιστάμενον)" (Graham 2010, 213). Finally, note that "Light" is not the conventional opposite of "Night," which draws attention further to the otherness of the world.

34. These adjectives can be taken adverbially, which would entail that the notion of continuum is not invoked here (cf. McKirahan 2008, 196–99). However, given that Fragment 8 is replete with continua, I take Parmenides to be reinforcing the notion of a continuum.

35. The goddess anticipates these references to continuous indefinite magnitudes when she says that what-is is "continuous" (συνεχές), which I read as deliberately ambiguous. McKirahan is especially intent on showing that συνεχές

should be read in light of the role played by συνέχεσθαι at 8.23 (McKirahan 2008, 224 n. 39). I agree that συνέχεσθαι is key to the argument and that we are being asked to see that "what-is draws near to what-is" conceptually, not spatially nor in terms of some other indefinite continuum. However, whereas McKirahan feels that it is necessary to choose between spatial and conceptual senses of συνεχές, which, as he notes, runs into trouble with the adjectival sense of ξυνεχές at 8.25, my reading avoids this problem by insisting on a theoretical role played by indefinite spatial and temporal continua.

36. Cosgrove (2014, 15) denies any relationship between τό ἐόν and the world mortals name.

37. Against my view, Johansen (2016, 15 n. 45) argues that Parmenides would not use the language of "likeness" if we were not meant to take the spatial image in some sense metaphorically or nonliterally.

Section III

Doxa

Chapter 10

Parmenides's Doxa and the Norms of Inquiry
A Case Study of the Fragments on Astronomy

SOSSEH ASSATURIAN

The poem of Parmenides falls into three parts. In the Proem, the goddess announces that she will tell the kouros two things: (1) "the unshaken heart of well-persuasive truth," and (2) "the opinions of mortals." The rest of the extant poem is occupied by these two topics. The Truth from Fragments 2 to 8.50 contains a series of deductions, which yield the signs (σήματα) of what-is as something unchanging, ungenerated, imperishable, a whole of a single kind, and complete. The remaining part of the poem from Fragment 8.51 to Fragment 19 constitutes the Doxa and outlines the dualistic cosmology underlying the changing, multifarious, perishable world of sense experience, which by its very nature lacks the features of what-is and becomes characterized by the goddess as "untrustworthy" and "deceptive."

For this reason and others, much of the scholarly sentiment regarding the Doxa has traditionally been disparaging of its philosophical value, relegating it to a mere dialectical exercise, a warning, or a "negative ideal model" for inquiry that illustrates the best possible cosmological model that nevertheless still falls short of truth.[1] As a result, the Truth has garnered focus as the philosophically significant section of the poem. But given its length relative to the whole poem, as well as the accuracy and ingenuity of some of the observations recorded in the Doxa, this part of the poem demands the reader's attention. Given this, it is difficult to believe it was written to serve a wholly negative purpose. There is, accordingly, an emerging appreciation for the scientific innovation evidenced by the Doxa's descriptions of phenomena belonging to the sensible world. This raises a critical question about whether, in dismissing the Doxa, we are missing something important about the philosophical contribution of the poem and about Parmenides's role in the history of philosophy and science. My aim in this chapter is to explore and motivate the possibility of an interpretation that treats this part of the poem as philosophically interesting in its own right.[2]

First, I set aside the question about the ontological status of Doxastic things and provide some considerations for why the Doxa deserves our serious attention by focusing on Fragments 10, 14, and 15: the fragments on astronomy. Next, I suggest a more positive place for the Doxa within the context of the poem as a whole through a discussion of the poem's embedded notion of inquiry as a goal-directed activity governed by domain-specific norms that are dictated by features of the objects proper to each kind of inquiry. I suggest that reading the poem as carving out these norms challenges some of the traditional interpretive assumptions about the status of the Doxa. Here, I contrast the purely "rationalist" norms that govern inquiry into what-is—the essences or natures of things—in the Truth with the contents of the Doxa, which appear to be the results of rigorous, principled empirical investigation into the world of sense experience.[3] Each type of inquiry yields epistemic improvement for the inquirer, so long as she correctly identifies the domain of her inquiry and proceeds in adherence to the zetetic norms proper to that domain. The "deceptiveness" of inquiry into Doxastic things is not, I argue, an inherent feature of inquiry in this domain, but the consequence of misinterpreting the domain to which the object of inquiry belongs and thus of violating Parmenides's general zetetic principle, with the wrong expectations about the kind of information the inquiry will produce. This, I suggest, constitutes Parmenides's diagnosis of the errors of the Ionian natural philosophers: they attempt to answer questions about fundamental reality and first principles with methods that are proper to investigating only the mechanics of the changing, contingent, sensible world.

On my reading, far from serving as a mere dialectical exercise, warning, or "negative ideal model" that outlines the illusory world of appearances, the Doxa demonstrates that the sensible world is a legitimate domain of inquiry that, when successful, produces empirical knowledge that contributes to the inquirer's epistemic improvement. The Doxa's role in the overall poem is therefore entirely complementary to the Truth. Together, they constitute a demonstration of what there is to know and how, as responsible inquirers, we can go about acquiring knowledge.

In Defense of the Doxa: The Fragments on Astronomy

There are at least two important considerations against interpretations of the Doxa that minimize its philosophical significance in the poem.[4] One is the extensive length of the Doxa relative to the other parts of the poem. A second is its content. Specifically, it contains a detailed description of a diverse scope of scientific theories. In this section, I take each consideration in turn and show how together they provide background motivation for reading the Doxa as a section of the poem that offers a positive philosophical contribution and deserves our serious attention.

First, while the Truth survives in near entirety at approximately seventy-eight lines, it is generally thought that the surviving text of the Doxa, which constitutes roughly a third of the extant poem at forty-four lines, is only about a tenth of what one would have found in the original poem.[5] This means that in comparison to the Truth, the Doxa comprised a much more substantial proportion of the original poem. We might say, then, that the Doxa already demands our attention due to its relative length. For dismissing it is tantamount to dismissing most of the poem, raising a question about why Parmenides would write so much about a topic that he ultimately did not believe requires the serious attention of the reader.

Second is the content of the Doxa. In this part of the poem, the goddess sketches a cosmology composed of the interaction between light and night, which contains a detailed exposition of various theories pertaining to phenomena that belong to the sensible world, including topics in astronomy, geography, theogony, and embryology. Much of the recent work on the Doxa has focused on the scientific ingenuity evidenced by its contents. Here, I take the fragments on astronomy as a case study of the systematicity and scientific rigor that underlie Doxa's accounts. Later, I use these findings to discuss the methodology behind the hypotheses in Doxa.

Of the Doxa's fragments on astronomy, Fragments 10, 14, and 15 stand out as a set of texts that have been embraced by both Parmenides scholars and specialists in the history of science. It is worth considering them in depth.

εἴσῃ δ'αἰθερίαν τε φύσιν τά τ' ἐν αἰθέρι πάντα
σήματα καὶ καθαρᾶς εὐαγέος ἠελίοιο
λαμπάδος ἔργ' ἀίδηλα καὶ ὁππόθεν ἐξεγένοντο,
ἔργα τε κύκλωπος πεύσῃ περίφοιτα σελήνης
καὶ φύσιν . . .

[You shall know the nature of the aether and all
the signs in the aether, and the destructive works of the
    shining sun's
pure torch and whence they came to be,
and you shall learn the wandering works of the round-faced
    moon
and its nature . . .][6] (10.1–5)

νυκτὶ φάος περὶ γαῖαν ἀλώμενον ἀλλότριον φῶς . . .
[A light by night, wandering around earth with borrowed
    light[7] . . .] (14)

. . . αἰεὶ παπταίνουσα πρὸς αὐγὰς ἠελίοιο.
[. . . always gazing toward the rays of the sun.](15)

Taken together, these texts contain evidence for several striking insights about the behavior of various celestial bodies. First is the observation about the source of the moon's light. Fragments 14 and 15 together constitute a treatment of the sun and the moon in relation to each other. In Fragment 14, we find two observations about the moon: first, it wanders or revolves around the earth, and second, it is not self-illuminating, and "borrows" its light from some other source. Fragment 15 explains that the source of lunar light is the sun, because the light side of the moon is always facing the sun, reflecting the sun's light. Fragment 10 further supports this hypothesis about the source of the moon's light. For first, it describes the sun as a "shining," "pure torch": as something that emanates light, thereby lighting other things. It also repeats the characterization of the moon as something that "wanders," or as something that moves. If we can understand the moon as a moving thing, we can hypothesize that the movements of the moon are responsible for lunar phases. By tracking lunar phases, we observe that the side facing the sun is always the side that is luminous. Fragments 10 and 15 therefore provide empirical evidence for the claim at Fragment 14. The three fragments together ascribe heliophotism (or solar illumination) to the moon and explain that lunar phases are caused by the reflection of the sun's light on the moon.

As far as primary evidence is concerned, Fragments 14 and 15 are sometimes cited as the earliest securely attested record for the discovery of heliophotism. But heliophotism also appears in testimonia as a hypothesis held by Thales, Pythagoras, Anaxagoras, and others. As a result, some believe that the discovery of heliophotism precedes Parmenides, and that it was merely recorded by him in the Doxa.[8] However, Daniel Graham has convincingly argued through a critical survey of the testimonia attributing this view to earlier philosophers that heliophotism must in fact be a Parmenidean discovery.[9] Indeed, while there is no consensus on this question, scholarly sentiment seems to be slowly shifting toward a view of Parmenides as the discoverer of various astronomical phenomena, including those that appear in the Doxa.[10]

Alexander Mourelatos argues that Fragment 10 explains an additional phenomenon: its mention of the "destructive works of the shining sun's pure torch" is a reference to how the sun's glare dims certain bands of the sky as it moves along its annual circuit. In other words, Fragment 10 tells us that the presence and absence of the sun's glare explains the disappearance and reappearance of constellations and stars. This includes the alternating disappearance and reappearance of the Morning Star and the Evening Star, which Mourelatos points out is inferentially connected to the realization that the Morning Star and the Evening Star are the same celestial entity: a discovery attributed to Parmenides by Aëtius in Fragment A40a.[11]

Indeed, many of the significant astronomical discoveries that are attributed to Parmenides in testimonia do not appear explicitly in the Doxa. At Fragments A1 and A44, for example, Diogenes Laertius attributes to Parmenides the view that Earth is spherical, while Fragments A37 and A44 outline Parmenides's geocentric model of the cosmos.[12]

But if we are concerned with whether the Doxa is worthy of serious philosophical consideration, why should we worry about the testimonia? While the discoveries and views attributed to Parmenides in the testimonia do not explicitly appear in the surviving fragments of the Doxa, they are nevertheless important pieces of the puzzle. This is not only because, as Mourelatos shows, some of the views that are attributed to Parmenides in the testimonia are inferentially relevant to what we find in the fragments of the Doxa, but because they help paint a picture of Parmenides as a rigorous scientist concerned with using observable data to support his hypotheses and speculative theories about the mechanics of the world of senses.

Putting together these considerations, it is difficult to believe that Parmenides would compose a lengthy, highly detailed, and innovative exposition of the results of what was clearly a rigorous investigation if he believed that the realm of this investigation was in fact *not* worthy of analysis, and further, that his reasons for developing it would be merely to warn his readers, or to provide an example of a metaphysical confusion or a negative ideal model.[13] Neglecting the Doxa as a significant philosophical and scientific achievement therefore hinders our full appreciation of Parmenides as a revolutionary figure in the history of science. Importantly, in doing so, we risk missing something vital about the poem, as I hope the rest of this chapter shows.

The World of Sense Experience as a Domain for Inquiry

Discussions of Parmenides's Doxa typically focus on questions about the ontological status of Doxastic things relative to the ontological status of what-is. Shaul Tor points out that the focus on this question yields the traditional response that Doxastic things belong to the illusory world of appearance and sense experience, in sharp contrast to the fundamental reality of what-is. He argues, in addition, that this characterization of Doxastic things has resulted in a negative view of the Doxa's overall contribution.[14] In the previous section, I set aside questions about the ontological status of Doxastic things and suggested that certain features of the Doxa merit an interpretation of the poem that is predicated on a different assumption: the Doxa makes a positive contribution that warrants serious attention. In this section, my goal is to suggest that Doxastic things are a legitimate domain for inquiry that, when successful, produces empirical knowledge.[15]

To start, the poem's focus on the theme of inquiry and knowledge is well attested in the literature. The poem's theme is summarized as follows by Charles Kahn: "As perceptive readers have always seen, the narrative of the journey to the goddess is an allegorical representation of Parmenides' enterprise as a quest for knowledge . . . The problem Parmenides raises from the beginning of his poem is not the problem of cosmology, but the problem of knowledge, more exactly, the problem of the search for knowledge."[16] Indeed, the poem begins with the goddess's comment at the end of the Proem, in which she welcomes the young man by claiming that she will guide him on a journey into two domains for epistemic exploration: the elements of Truth and the opinions (δόξας) of mortals.

This theme is further pronounced in the subsequent fragment, where the goddess tells the kouros about "the only routes of inquiry (διζήσιός) that are for knowing (νοῆσαι)."[17] This comment narrows the scope of the goddess's exposition. For it is not an account of inquiry simpliciter that she shares with the kouros. Rather, it is a particular kind of inquiry: that is, inquiry that, when successful, produces knowledge. Thus, the goddess's references to inquiry throughout the rest of the poem are references to a goal-directed activity, where the goal is knowledge. The language of "routes" of inquiry is metaphorical for ways of investigating, and the ways of investigating are, as we shall see, distinguished by the domains for investigation. In what follows, we should therefore expect to find out about the domains of inquiry that will help us achieve the goal of attaining knowledge. These ways, she explains, are the route of inquiry into what-is and the route of inquiry into what-is-not. By the end of Fragment 2, then, there appear to be two domains for inquiry, which are distinguished by their respective objects: what-is and what-is-not.

At this point, in establishing whether the world of the Doxa counts as a domain of inquiry, we confront the notoriously difficult question about the number of routes for inquiry endorsed by the goddess. Because this question is discussed at length elsewhere this volume, here I will briefly summarize some of the big-picture textual problems and endorse the three-route orthodoxy.[18]

The confusion about the number of routes for inquiry arises because two routes (the routes of what-is and what-is-not) are introduced in Fragment 2, while at Fragment 6.3, the goddess describes a "backward-turning" route in which "what-is and what-is-not are thought to be the same and not the same."[19] Historically, two-route readings have read the goddess's comment at Fragment 2 as exhaustive, and they have identified the two routes as the route of what-is and the route of what-is-not, while three-route readings have treated 6.3's "backward turning" path as a separate, "mixed" third route whose objects are what-is-and-is-not, including Doxastic things.[20]

Two-route readings have the advantage that the interpretation fits with the poem's overall theme of duality (for example, between what-is and what-is-not, Truth and Doxa, Light and Night). But traditional two-route readings have trouble paying the Doxa its dues, and in relegating Doxastic things to the route of what-is-not, they fail to take into sufficient account the differences in how the route of what-is-not and the route of Doxastic things are each described. For instance, it is emphasized throughout the poem that the route of Doxastic things is a "mixed" path of is-and-is-not, while the route of what-is-not is always simply the route of what-is-not *as such*: that is, it is not described as mixed with what-is. In addition, the goddess prohibits inquiry along the route of what-is-not, as a "path entirely unable to be investigated" for the reason that one can neither "know (γνοίης) what-is-not" nor can one "point it out (φράσαις)."

But the Doxa arrives to us as an account of things we can point out in the world around us, so that its contents, in some straightforward sense, *are* identifiable. We can, for instance, reliably indicate the moon or the sun. Second, by the goddess's own lights, Doxastic things are attached to some kind of epistemic value. This is indicated throughout the poem by the presence of various knowledge and learning verbs, the objects of which are Doxastic things. At the end of Fragment 1, for example, the goddess claims that one must learn (πυθέσθαι) both matters pertaining to Truth, as well as the opinions (δόξας) of mortals. Verbs pertaining to learning or knowledge appear throughout the Doxa as well. At 8.51, the goddess asserts, "From here on, learn (μάνθανε) mortal opinions . . . ," attaching one such verb to the contents of the Doxa, which appear subsequent to these instructions. These verbs also appear in the astronomy fragment, Fragment 10, quoted above. There, the nature and signs of the aether are both objects of the knowledge verb είσηι. Likewise, the nature and movements of the moon are the objects of the learning verb πεύσηι. That Doxastic things are the objects of learning verbs suggests that we can learn about them, which suggests that we can, indeed, engage in inquiry about them. By the same token, that they are objects of knowledge verbs suggests that we can know them. The way Doxastic things are described therefore explicitly counts against treating them simply as what-is-not *as such*.[21] Two-route readings that relegate Doxastic things to the route of what-is-not, then, cannot offer a fully satisfactory account of the Doxa or of the differences in the language the goddess employs throughout the poem to describe Doxastic things in contrast to what-is-not.

Three-route readings fare substantially better on this score. By treating the route of Doxastic things as separate from the route of what-is-not, the differences between how the two routes are characterized is fully taken into account, leaving open the possibility that there might be a positive place for the Doxa.[22]

However, as we have seen, when the goddess describes the second route in Fragment 2—that is, the route of what-is-not—we are forbidden from inquiring into it. For she claims that it is a route "entirely unable to be investigated" due to its inconceivability. The practical impossibility of inquiry into what-is-not also turns up at 6.1–2, where the goddess explains, "It is right both to say (λέγειν) and to think (νοεῖν) that it is what-is: for it can be, but nothing is not." Here, again, the binary between what-is and what-is-not is brought to our attention. The possibility of inquiry into what-is is framed as an ability to speak and think about it. If we add to this the qualification that the kind of inquiry under discussion is an inquiry that produces knowledge, we can see that inquiry into what-is-not is impossible not only because we cannot speak, think about, or indicate nothing at all, but also because inquiring into nothing would not produce any kind of epistemic improvement. We might say, because inquiry along this route is impossible, that it is not a genuine route at all. Indeed, this is exactly what the goddess says:

. . . κέκριται δ' οὖν, ὥσπερ ἀνάγκη,
τὴν μὲν ἐᾶν ἀνόητον ἀνώνυμον, οὐ γὰρ ἀληθής
ἔστιν ὁδός, τὴν δ' ὥστε πέλειν καὶ ἐτήτυμον εἶναι.

[. . . and it has been decided, as is necessary,
to leave the one [route] inconceivable and unnamed, for it is
   not a true
route, so that the other [route] is and is genuine.] (8.16–18)

Importantly, the claim that the route of what-is-not is not a genuine route is not equivalent to the claim that it is not a route at all. For the latter of these two claims would have the goddess contradicting Fragment 2's claim about there being two routes, of which what-is-not is one. To avoid this, we can take the route of what-is-not as merely a nominal route meant to establish the conceptual space for inquiry through a formal consideration of what-is and its negation. For a route to be genuine, we must be able to speak and think about it in a way that produces knowledge. Since the route of what-is-not turns out to be a nonstarter, the goddess further explores the conceptual space between what-is and what-is-not by offering an additional route for inquiry that falls in between the two as an intermediate. We can read Fragment 6 as a programmatic comment about how the rest of the poem proceeds: that is, from the first route of inquiry (the route of what-is), and from there, to the "backward-turning route" of mortals. The rest of the poem adheres to this signpost as she describes the characteristics of what-is through a series of deductions in Fragment 8 and directly after, moves to her exposition of Doxastic things.[23]

Given that we have good reasons to separate Doxastic things from what-is-not, this suggests that the mixed, "backward-turning" route of mortals is not the route of what-is-not. Rather, it is a new, genuine route introduced by the goddess at 6.3, which fits into the conceptual space between what-is and what-is-not established in Fragment 2 by mixing what-is with what-is-not. This is to say that Doxastic things combine elements of both what-is and what-is-not, given their nature as changing, multifarious entities, processes, and phenomena that are at the same time something (but do not exhibit the signs of what-is in Fragment 8) and not nothing (by virtue of being available for thought and speech). The goddess's statement in Fragment 2 about stating the only routes for inquiry, then, is not completed until 6.3, when the mixed route of Doxastic things is finally offered as a domain for inquiry.

We have, then, three routes for inquiry: the route of what-is, the route of what-is-not, and the route of what-is-and-is-not. While the route of what-is-not is only a nominal route, the route of what-is and the route of what-is-and-is-not are genuine routes. This means that their objects are available for thought and speech, and if successful, inquiry along these two routes produces knowledge. In the next section, I suggest that the poem as a whole is an exposition of zetetic norms, where the kind of inquiry under consideration is inquiry conceived as a goal-directed activity with norms that are domain-specific. The world described in the Doxa is one such zetetic domain with its own set of zetetic norms.

Two Genuine Domains of Inquiry, Two Sets of Zetetic Norms

Recent work in contemporary epistemology has focused on inquiry, and in particular, on the norms of inquiry. The idea, roughly, is that in the pursuit of information, there seem to be certain rules about how one ought to go about obtaining the information one seeks. This is a question about the norms one should conform to when inquiring. Say, for example, that I want to know whether it is currently raining outside in my location. To find out, I can go outside, look through the window, or check the current weather on my phone. These all conform to a norm of inquiry: if my goal is to find the answer to some question (e.g., whether it is currently raining), then I should do what is required to get the answer to that question (e.g., look outside the window). This zetetic norm is a basic norm of instrumental rationality.[24] But there are, of course, other norms of inquiry.[25] We might say that if our inquiry is not governed by certain zetetic norms, there is a good chance that the goal of our inquiry—whether it is knowledge or something else—will not be achieved.[26]

In the previous section, I suggested that Parmenides's goddess is interested in inquiry as a goal-directed activity that aims at knowledge and that she offers two genuine routes for knowledge-producing inquiry: the route of inquiry into what-is and the route of inquiry into what-is-and-is-not (or into Doxastic things). It remains to be seen what inquiry in these two domains in fact entails. In this section, I suggest that in Fragment 8, the goddess offers an example of inquiry into what-is, while in the Doxa, she offers examples of inquiry into what-is-and-is-not. In her demonstrations, we find zetetic norms that turn out to be domain-specific.

In Fragment 8, the goddess follows through on her promise to guide the kouros along inquiry into what-is and leads him through a series of deductions, unraveling the signs that point to its characteristics. The distinctiveness and rigor of the rationalist methodology that appears in this fragment has been noted.[27] That this fragment constitutes an a priori series of arguments that are propelled not by the evidence or justification provided by sense experience, but by reason alone, is rooted in both the nature of the arguments themselves and the language the goddess uses to introduce the arguments.

With respect to the significance of the goddess's language, at 7.4, directly before guiding the kouros along inquiry into what-is, she warns, "judge (κρῖναι) by reasoning (λόγῳ) the examination spoken by me." Two things are noteworthy about this. First, since this line precedes the account of inquiry into what-is, we can safely assume that "the examination" in question is inquiry into what-is. Second, it tells us that the method for inquiry into what-is must proceed by reason, that is, not by sense experience. Importantly, this instruction is given in the imperative, which we can read as having normative force. Because it appears as part of a directive, the idea is effectively that the kouros indeed *must* or *should* use reason—and not "an aimless eye" or "an echoing ear"—to judge the arguments that reveal the characteristics of what-is. This constitutes a zetetic norm for inquiry in this domain (call it ZP_1 for "zetetic principle 1"):

ZP$_1$: Inquiry in the domain of what-is must be governed by reason and cannot proceed by the evidence of the senses.

The arguments of Fragment 8 follow ZP$_1$. Take, for example, the statement by the goddess at 8.13–16 that "the decision about these things is in this: is or is not." As Michael Wedin argues, the claim in these lines is that the justification for whether what-is is generated and perishing must be based on "is or is not." In other words, the arguments about what-is must be governed by the law of the excluded middle, an a priori premise that was introduced in Fragment 2.[28] The arguments of Fragment 8 in fact produce at least four deductive consequences that take principles stated in Fragment 2, Fragment 3, and Fragment 6 as assumptions.

For the sake of simplicity, let's take the argument for the conclusion that what-is is indivisible and continuous as an example of inquiry into what-is, as it is demonstrated throughout Fragment 8. This will be sufficient as an instance of ZP$_1$ at work in the Truth. The text reads:

οὐδὲ διαιρετόν ἐστιν, ἐπεὶ πᾶν ἐστιν ὁμοῖον·
οὐδέ τι τῇ μᾶλλον, τό κεν εἴργοι μιν συνέχεσθαι,
οὐδέ τι χειρότερον, πᾶν δ' ἔμπλεόν ἐστιν ἐόντος.
τῷ ξυνεχὲς πᾶν ἐστιν· ἐὸν γὰρ ἐόντι πελάζει.

[Nor is it divisible, since it is all alike,
and not at all more here and less there, which would keep it
   from holding together,
but it is all full of what-is.
Therefore, it is all continuous; for what-is draws near to
   what is.] (8.22–25)

The argument for the continuity of what-is runs as follows. First, something's not being more here and less there is equivalent to it being full of what-is. If it does turn out that something is more here and less there, or that it is not full of what-is (i.e., if either of the two claims in the previous premise fails), this would prevent what-is from holding together. But since neither of those two claims fails, what-is draws near to what-is. This is just what it means for what-is to be continuous. Therefore, what-is is continuous. The argument for the indivisibility of what-is appears as a companion to this. Based on 8.22, we get the following conditional: if what-is is all alike, then it is not the case that what-is is divided. Equivalently, if what-is is divided, then it is not the case that what-is is all alike. From the assumption that what-is is divided, if we accept that what-is has parts, and that those parts are not "all alike" (i.e., that they are different) then it would be possible for what-is-not to be. But since we know that what-is-not cannot be, what-is either has no parts or its parts are identical to itself. If it has no parts, it is indivisible. If its parts are identical to itself, they are all alike and do not count as true parts, which entails that what-is is indivisible. Therefore, what-is is indivisible.[29]

Constructing this argument and determining the truth of its premises does not require perceptual evidence or verification from the empirical world. For in 4.1, the characteristics of what-is are "absent" and yet "securely present to the mind." They are absent insofar as one cannot perceive them with the senses. We cannot touch the imperishability of what-is with our hands or see its completeness with our eyes, for example. But we can grasp those features of what-is with the mind, through reasoning. The proper method for inquiring into what-is is determined by the characteristics of what-is and our means of grasping those characteristics. Because the characteristics of what-is are imperceptible, we cannot use perception to grasp them: hence, ZP_1. We can therefore read the argument for the continuity and indivisibility of what-is as an example of the goddess demonstrating inquiry into what-is by using ZP_1. Indeed, additional examples of ZP_1 are in operation throughout Fragment 8, where the goddess establishes that what-is is also ungenerated and imperishable (8.5–21), motionless (8.26–31), and complete (8.32–49).

This is in sharp contrast to what we read in the Doxa. Earlier, I explained that Fragment 15 and elements of Fragment 10 act as evidence for the conclusion in Fragment 14. There, the evidence for heliophotism is that (1) the moon moves around earth, (2) as the moon moves around earth, we can observe different lunar phases, or differences in the amount of light the moon receives, (3) the illuminated part of the moon is always facing the sun, and (4) the amount of light the moon receives depends on its position relative to the sun. Claims (1) through (4) are grounded in perceptual, observable evidence of the behavior and appearance of the moon relative to the sun. If we wanted to verify these claims, there would be no way to do so from the armchair, so to speak. Rather, we must rely on sensory data collected from our observations of the moon and the sun.

By comparing inquiry into what-is in Fragment 8 with inquiry into what-is-and-is-not in the Doxa, we can observe the emergence of distinct norms of inquiry. In particular, we find a different zetetic norm at work in the world of Doxastic things than in the domain of what-is. Call it "ZP_2" for "zetetic principle 2":

> ZP_2: Inquiry in the domain of what-is-and-is-not must be governed by sense experience (and cannot proceed by reason alone).

Earlier, I explained that the characteristics of what-is, along with our means for grasping those characteristics, determine ZP_1. Similarly, because the characteristics of Doxastic things are available to sense experience and cannot be grasped or verified through reason alone, our perceptual capacities are the appropriate means for grasping them. This determines ZP_2.

The fragments on astronomy serve as an appropriate case study for ZP_2 in practice, given the accuracy of the hypotheses, as well as the fact that the kinds of observations (e.g., of the lunar phases) that would ground the claims in those texts are clear. However, it is not immediately obvious what observations would ground some of the other hypotheses that appear in the Doxa. For instance, nothing in the extant fragments of the Doxa seems to serve as evidence for Fragment 17's hypothesis that males are conceived in the right part of the uterus and females in the left.[30] Indeed, while this bizarre-sounding hypothesis might strike us as an example of mistaken ancient "empirical" theorizing, if our thesis is that inquiry into Doxastic things is governed by sense experience, then it is worth exploring what observations might have justified it.

A testimony on Parmenides from Aristotle at *De partibus animalium* ii.2.648a29–31 helps us with this. There, Aristotle attests that according to Parmenides, females are warmer than males, on the grounds that heat is determined by an abundance of blood. At *De generatione animalium* iv.1.765b17–28, Aristotle claims that "Anaxagoras and others" hypothesized that females are hotter than males because females have more blood than males, and that the claim about females having more blood than males is tied to the observation that females menstruate.[31] Since the left is supposed to be hot, and the right is supposed to be cool, the idea is that the warmer sex is conceived in the left part of the uterus, while the cooler one is conceived in the right part.[32] Hence, even the bizarre hypothesis in Fragment 17, incorrect as it may be, is governed by ZP_2.

We can therefore generalize from the demonstrations of inquiry into what-is in the Truth and of inquiry into what-is-and-is-not in the Doxa, a general principle that guides inquiry for Parmenides; call it the "general zetetic principle" (GZP):

> GZP: Inquiry must be governed by the norms proper to the domain to which the object under investigation belongs.

As we have seen, what underlies this principle for Parmenides is that both what-is and Doxastic things are available as objects of inquiry. Because they have different characteristics (the former is not accessible by sense perception, while the latter are not accessible by reason alone), successful inquiry into each type of entity will require the inquirer to proceed by sets of norms that are determined by the object under inquiry. If the inquiry is successful, accounts of both types of things yield epistemic improvement for the inquirer.

Positive interpretations of the Doxa, however, must contend with the fact that the accounts in the Doxa are purportedly disparaged by the goddess herself as "deceptive" (8.51), while the route of is-and-is-not is described as the "backward-turning" path (6.9) in which there is "no true trust" (1.30). It is also characterized as the path of helpless, confused mortals who "know nothing" (6.4–5). Taken at face value, these negative descriptions of the Doxa count explicitly in favor of negative readings. However, if I am right that Parmenides is committed to something like GZP, and that he accepts Doxastic things as legitimate objects for an epistemically valuable type of inquiry, then an alternative reading of these texts would neutralize this evidence as explaining what would happen if an inquirer violates GZP in the process of inquiring.

One way to violate the norm would be to mischaracterize the object of inquiry, and therefore to attempt an inquiry guided by the wrong set of norms. For example, I might mistake inquiry into the source of the moon's light for inquiry into what-is. I might then try to discover facts about lunar phases in the same way in which I would discover facts about what-is: that is, without observing them at all, and by relying on my commitment to certain a priori principles to tell me why the moon appears to change. In this case, my inquiry would fail on the grounds that it was not guided by the proper norm; reasoning from the law of noncontradiction, for instance, will not tell me what I want to know about the moon. My inquiry would fail for the same reason if I tried to investigate what-is by observing the perceptible world around me. In both scenarios, the inquiry fails because using the wrong method of investigation either does not yield an account at all or it produces an account that is incorrect and therefore does not contribute to the inquirer's epistemic improvement. However, note that the risk here is not merely a matter of one's inquiry failing because it is guided by the wrong set of norms: it's that one has inquired using the wrong set of norms *because* one has misinterpreted inquiry into a Doxastic thing for inquiry into what-is (or vice versa).[33] The force of GZP is therefore that it is crucial for the inquirer to identify the domain of inquiry correctly. In other words, even if I can recognize that my inquiry into lunar phases ought to be guided by my observations of the moon, the account produced by that inquiry only has epistemic value if I acknowledge that lunar phases are explicitly a Doxastic thing. Lunar phases do not count as what-is, or as fundamental reality. The goddess's insistence on mortal error highlights precisely this: the human tendency to misinterpret inquiry into Doxastic things as inquiry into what-is, and vice versa. We can therefore add a further refinement to GZP:

> GZP *: Inquiry must be governed by the norms proper to the domain to which the object under investigation belongs **and** both the object of inquiry and the account produced by the inquiry must be accurately construed by the inquirer as belonging to that domain.

The "deceptiveness" of inquiry into Doxastic things is therefore not an inherent feature of inquiry in this domain. Indeed, the epistemic value of inquiry into what-is-and-is-not is emphasized at 1.3: "Nevertheless, you will learn these [accounts of Doxastic things] too, how it were right that the things that seem be reliably, being indeed the whole of things." Doxastic things, in addition to what-is, are after all a part of the "whole of things." We mortals live in a world that we navigate using our senses. As such, Doxastic things are available to us as objects for investigation and as items and processes that we can, in some sense, reliably come to know. Their deceptiveness is the result of our own "two-headedness," of the fact that as mortal inquirers, we are capable of knowing things both through abstract reasoning and through empirical theorizing. The goddess's commentary on mortal error in Fragment 6 amounts to this: if we are not careful, we can drive ourselves into a kind of epistemic "helplessness" that renders us unable to keep our minds from "wandering," given our capacity to engage in different kinds of inquiry.

Some Lessons

At the beginning of this chapter, I argued that negative readings of the Doxa do not offer a satisfying explanation of its role in the poem of Parmenides. I therefore set out in search of an interpretation of the Doxa that adequately accounts for both its substantial length relative to the poem as a whole, as well as the accuracy and highly detailed nature of its descriptions of Doxastic things. To do so, I used philological and conceptual clues in the poem to establish that the route of Doxastic things (what-is-and-is-not) constitutes a route of inquiry for knowing, as the 'mixed' path of mortals, distinct from both the route of what-is-not and the route of what-is. Using the fragments on astronomy, as well as the arguments in Fragment 8 for the continuity and indivisibility of what-is, I showed that throughout the course of the poem, the goddess demonstrates that inquiry is governed by domain-specific zetetic norms, which are subsumed under a refined version of what I called the "general zetetic principle." According to this principle, inquiry must be governed by the norms proper to the domain to which the object under investigation belongs, and both the object of inquiry and the account produced by the inquiry must be accurately construed by the inquirer as belonging to that domain. The norms of inquiry for Parmenides are domain-specific because the objects belonging to the two domains for inquiry have different characteristics and are therefore available to mortal inquirers by different means: either through reasoning or through empirical observation. Deception and confusion are therefore not inherent to inquiry into Doxastic things. Rather, the goddess's warnings about inquiry along this "mixed" path are grounded in the mortal habit of violating the norms of inquiry, in mistaking accounts of Doxastic things for an account of what-is. This habit is what causes inquiry to fail and leads mortals away from the goal of epistemic improvement either in the form of empirical knowledge, or of knowledge of what-is.

To close, I suggest that the joints at which Parmenides carves his views of inquiry and the views of mortal error that are presented in the poem are both targeted at the Ionian tradition that precedes him. While Parmenides accepts that the natural world is a legitimate domain for inquiry that, if successful, can produce empirical knowledge, he warns us that if, as inquirers, we seek to identify fundamental reality or first principles, we cannot successfully do so by investigating the perceptible world or by following norms proper to inquiry into the world of sense experience to guide our investigation. It is precisely the confusion of

Doxastic things for what-is that led the Ionians to posit water, air, or a limitless mass as first principles. Their conclusions about first principles are reached as a result of erroneously inquiring into what-is by looking to the perceptible world, instead of by using reason alone to derive a priori principles through which successful investigation into what-is must proceed. Parmenides is therefore both a champion of the sensible world as an acceptable source for knowledge, and a critic of the Ionian natural philosophers who commit the category mistake of positing that the very same elements of the perceptible world *are* what-is (or at the very least, that they provide important insights into what-is). The Doxa, then, is a significant contribution in its own right, not only because it is a dossier of the achievements of Parmenides as a rigorous, principled scientist, but because it is also a glimpse into Parmenides's sophisticated philosophical meditations on how to inquire successfully in any domain.

Notes

1. For views of the Doxa as a dialectical exercise, see Owen 1960 and Cosgrove 2014. For an interpretation of the Doxa as a warning, see Nehamas 2002. For views of the Doxa as a negative ideal model, see Long 1975, Barnes 1982, de Rijk 1983, Gallop 1984, Inwood 2001, Curd 2004, Warren 2007, Miller 2011, and Bryan 2012. See also DeLong's chapter in this volume for an interpretation that sets out to unify the three main parts and retains a disparaging view of the Doxa. Throughout, I refer to these interpretations of the Doxa as "negative" views. Other possibilities for negative interpretations are outlined at Tor 2015, 4. "Positive" interpretations are those that interpret the Doxa as making a substantial philosophical contribution beyond merely serving as a dialectical exercise, a warning, or a negative ideal model.

2. I call the items, phenomena, and processes that belong to the world described in the Doxa (i.e., those lacking Fragment 8's signs) 'Doxastic things' for brevity, following Tor 2015.

3. In her chapter in this volume, Bryan challenges the view of Truth as operating from rationalist norms, focusing on the role of divine revelation in uncovering the signs of what-is.

4. Cf. Tor 2015, 4–5.

5. See Palmer 2009, 160 for estimates of the length of the original poem.

6. Translations are my own, in consultation with Curd 2011, 55–65 and Graham 2010, 203–42.

7. I read this fragment with the MSS νυκτὶ φάος in place of Scaliger's νυκτιφαές. For a defense of the MSS reading, see Mourelatos 2012.

8. See, e.g., Tarán 1965, 245–46; and Coxon 2009, 374–75. For the view that Fragments 14 and 15 do not imply heliophotism, see, e.g., Diels 1897, Heath 1912, and Gallop 1984, 85.

9. See Graham 2002.

10. On the testimonia's reliability specifically as pertaining to Parmenidean astronomy, see Graham 2013, 88–92 and Finkelberg 1986, 303–17, which point to Parmenides as the original discoverer of these phenomena. See Palmer 2009, 161–62 for a brief discussion of scholarly sentiment on the contents of the Doxa as Parmenides's own reflections and discoveries.

11. Mourelatos 2011. These observations are also discussed in Casertano 2011 and Cerri 2011.

12. The view of Earth as spherical is also attributed to Parmenides by Aëtius at *Dox.* 380, 13–18. See Kahn 1994, 80, and 234 for critical analysis of the nature of these testimonia. For responses to Kahn, see the discussions referenced in the two previous footnotes. For a discussion of possible tensions between Fragments A37 and A44, see Finkelberg 1986.

13. This latter point is defended in Tor 2015, 5.

14. See Tor 2015, 4.

15. Tor's reading is correct in its assumption that interpretations of Parmenides should take up the Doxa as an important part of the poem. The reading I suggest here is congenial to the one defended in Tor 2015, insofar as it takes Parmenides's epistemology to be the central lens through which the whole poem is interpreted. However, Tor's central thesis is that the two parts of the poem are examples of different kinds of knowledge: that is, divine knowledge in the Truth and mortal knowledge in the Doxa. My focus is not on different kinds of knowledge, but on Parmenides's notion of inquiry.

16. Kahn 1969, 704–5.

17. I translate νοῆσαι as "knowledge" to denote an epistemic success term. For a defense of this translation, see Kahn 1969, Coxon 2009, and Mourelatos 2008.

18. See Matthew Evans's chapter in this volume.

19. The manuscript of Fragment 6.3 contains a lacuna in a crucial part of the text in which the goddess claims, "For < . . . > you from this first route of inquiry, and then from that, on which mortals, knowing nothing, wander . . ." Diels supplies εἴργω so that the line reads "I forbid you," while Cordero (1979) and Nehamas (1981) supply forms of ἄρχω so that the line reads either, "I begin for you," or, "You will begin." On this debate, see Curd 2004, 51–63. See also Evans's chapter for a discussion of the implications of the various supplements for the number of routes.

20. Recent three-route interpretations are Palmer 2009 and Wedin 2014. For a survey of the historical precedent for three-route readings, see Smith 2020, which defends a two-route interpretation on which the route of what-is-not is not a route. Accounts of the two routes as the route of what-is and the route of what-is-not are Cordero 1979, Nehamas 1981, and Curd 2004.

21. Nor does the changing, perishable, non-eternal world described in the Doxa belong on the route of what-is, for it lacks the signs described in Fragment 8.

22. Though, of course, three-route readings still endorse a negative interpretation of the Doxa. See especially Wedin 2014 and Owen 1960.

23. On the connection between these deductions and the turn to the Doxa, see Eric Sanday's chapter in this volume.

24. This norm is articulated in Friedman 2020.

25. In the context of the history of philosophy, for example, it has been argued that for Aristotle, because animal generation is a single, goal-oriented process, inquiry into animal generation is governed by a zetetic norm that dictates that understanding the goal of the process of generation must be prior. See Lennox 2021, 88.

26. Throughout this chapter, I have been referring to the product of successful Parmenidean inquiry as either "knowledge" or "epistemic improvement" interchangeably to reflect the ambiguity in the various connotations of the different knowledge and learning verbs in the poem (since epistemic improvement can come in different forms, including knowledge, understanding, a more accurate credence, etc.) I do not take the use of different "epistemic" verbs to indicate a systematic, deliberate choice reflective of technical use on Parmenides's part. However, the goal of inquiry is a robust problem in contemporary epistemology. For a view of knowledge as the aim of inquiry (and a survey of other possibilities), see Kelp 2014, 227–32. For an epistemic-improvement view of the aim of inquiry, see Haas and Vogt 2020, 254–67.

27. Wedin 2014 represents a thorough appreciation of this by reconstructing a detailed version of the poem's deductions in formal logical notation.

28. For a detailed defense of this, see Wedin 2014, 9–11.

29. For the logic behind this argument, see Wedin 2014, 105–9.

30. That Fragment 17 describes the efficient cause of sex differentiation as the position of the conception of the fetus in the uterus is suggested in Lloyd 1966, 17 and 50. This reading of Fragment 17 has been challenged in Kember 1971, 70–79 on the grounds that the fragment can also be interpreted as hypothesizing that a fetus that is already male, for example, is conceived on the right side of the uterus (so that the fetus' position in the uterus is determined by its sex, and not the other way around). However, the context in which Fragment 17 is quoted in Galen's *Commentary on Book VI of Hippocrates's Epidemics* II.46 is a lengthier explanation not of how the side of the uterus the fetus is in is determined by the sex of the fetus, but of how the sex of the fetus is influenced by the side of the uterus it is in. The context of Fragment 17 therefore clarifies that the favorable interpretation of Fragment 17 is the one I assume here.

31. Parmenides is not named in the testimony at *GA*, but the reference to "others" is generally taken to include him. See Lloyd 1966 and the testimonia arrangement in Coxon 2009, 132–33.

32. See Mansfeld 2018 [2015], ¶ 15–17 on the directionality of hot and cold in Greek thought.

33. Cf. Tor 2015, 33–34, where mortal error is construed as a mix-up between knowledge of Doxastic things and divine knowledge (of what-is).

Chapter 11

The Essential Role of the Doxa in Parmenides's Teaching

JESSICA ELBERT DECKER

In this reading of Parmenides, the focus will be on the experiential aspects of his teaching and how they function in the text as a means of bringing his philosophical concepts into the lived experience of the student, emphasizing the importance of the doxa as the necessary experiential context in which alētheia can become intelligible. Parmenides's poem is a subtle, dense, and nuanced text best approached with some awareness of the poetic and religious conventions of its time.[1] When the poetic techniques used throughout the poem are made visible, the experiential nature of Parmenides's teaching becomes more accessible. Perhaps the most obvious of these techniques is Parmenides's use of Homeric allusions throughout his text; he frequently employs Homeric language and imagery to conjure up a plethora of associations for the listener.[2] Since the Homeric narratives were, for archaic Greeks, not read but performed, these Homeric associations evoke lively, visceral experience in the listeners.[3] When Parmenides invokes the chariot race in the *Iliad* (23.262–52), for example, ancient hearers would imagine it with rhapsodic intensity. To our contemporary ears, much of this instantaneous effect is lost. In particular, I will demonstrate that Parmenides is not, as some traditional readings claim, encouraging us to dismiss the senses in favor of "reason" but is instead offering an experiential teaching with precise techniques for becoming aware of certain habits of mortal thinking that obstruct sensation and perception.[4] Parmenides's teaching, while on the surface about the "ways of thinking," indicates the *limits* of mortal thinking and offers instruction on how to move beyond them. Thumos, often translated as "desire," begins Parmenides's journey in the Proem, not thinking or noos, and Parmenides's teaching offers the student guidance in cultivating both stillness and the ancient Greek power known as mētis, cunning resourcefulness. In the text, the goddess displays mētis especially through her cleverly deceptive use of double speak.[5]

Signposts on the Necessity of the Doxa

Two passages from Parmenides's poem will serve as introductory markers that express the necessity of the Doxa for understanding Parmenides's teaching. The first is lines 1.28–32, where the goddess explicitly tells the kouros, "What's needed is for you to learn *all things*: both the unshaken heart of persuasive truth and the opinions of mortals in which nothing can truthfully be trusted at all. And this too you will learn: how beliefs based on appearance ought to be believable as they travel all through all there is."[6] The second passage appears in the Doxa, where the goddess says, "I tell you the whole arrangement as it seems to mortals so that no mortal may ride on past you (παρελάσση) in knowledge" (8.60).[7] These two passages will be carefully explored to provide some orientation in understanding the role of the Doxa. The first statement makes very clear that knowing the "unshaken heart of persuasive truth" is not sufficient; the kouros must also learn the ways in which things appear to mortals because "beliefs based on appearance ought to be believable as they travel all through all there is" (1.28–30).[8] If the kouros is to learn "all things," he must learn to recognize that what appears to him as two opposite things—alētheia and doxa, for instance—is, from the perspective of the goddess speaking, a unity.[9]

The second passage invokes the famous chariot race in the *Iliad* with the term παρελάσση, which means "ride on past," as Antilochus does when he uses mētis to outwit his faster opponent.[10] This allusion suggests that the kouros, after learning both the doxa and alētheia from the goddess, will never be outwitted by his mortal peers: the teaching can give him a resourcefulness or cunning (qualities of mētis) uncommon among mortals. Most mortals are not only lacking experience of alētheia, but crucially, they also lack experience of the doxa since they fail to understand the habits of their own thinking. This criticism of mortals is expressed in very similar language in Heraclitus's Fragment 1, where he suggests that mortals are so completely unaware that they fail to recognize their own experience (they are "like those without experience") and cannot even discern between waking and sleep: "but other human beings are oblivious of what they do awake just as what they do asleep escapes them."[11] The grim and customary state of mortals is described in very explicit language in Parmenides's Fragment 6, where they are "carried along in a daze, deaf and blind at the same time." In this state, human beings are incapable of recognizing their own perception, which is binary and conditioned by both space and time; they are entirely unable to grasp the unity of alētheia.[12] Parmenides's teaching—understood with respect to both alētheia and doxa—can offer the student the necessary experience needed for understanding both the "is" and the manner in which it appears.

The Doxa portion of the poem is a demonstration of how things appear to mortals, and understanding this demonstration requires familiarity not only with the dense text of the Doxa but most crucially with the experience of alētheia, or more explicitly, with the experience of the "unshaken heart of persuasive truth." In Fragment 14, the goddess offers an image that can illuminate this relation between alētheia and doxa, using the light of the sun and the moon as demonstration. Alexander Mourelatos has called the description of the moon in Fragment 14 "the most successful line of poetry in Parmenides"[13] for its density and ambiguity: "Νυκτιφαὲς περὶ γαῖαν ἀλώμενον ἀλλότριον φῶς . . ." The first term, Νυκτιφαές, may be translated as "night shining" or, to emphasize the oxymoron, "nightbright"; ἀλώμενον conjures up images of wandering, as mortals in the third way have "wandering minds"; and the final phrase ἀλλότριον φῶς "is instantly recognizable as an imitation of the Homeric formula ἀλλότριος φώς, 'an alien man, a stranger.'"[14] The entire line may be translated as: "nightbright wandering round the earth a stranger's light."

The density and cleverness of this line cannot be overstated: the image and the language that informs it demonstrate how "beliefs based on appearance ought to be believable as they travel all through all there is": this passage produces an image of the doxa, of how mortal opinions appear. The oxymoron Νυκτιφαές (nightbright) is exemplary as a deceptive appearance: the moon appears to be shining, but its light is borrowed from the sun. The appearance that the moon is shining is real and is in fact the experience of any mortal who stares at the moon shining bright in the night; the truth adds another dimension to this as one who is "initiated" recognizes that the moon's light is reflected from the sun and not of its own origin. It is in this sense that "beliefs based on appearance ought to be believable": they are convincing, though they are not worthy of true pistis since there is another perspective that recognizes both the deceptive appearance of the night-shining moon *and* its borrowed light. Thus the doxa that mortals believe is a true *appearance* of what is; however, it is not a complete understanding because it sees only the surface and lacks depth. As Mourelatos points out, this line has further depths as the image of the moon in its round wanderings also invokes the Cyclops, naming him as a "wandering stranger."[15] These Homeric associations are significant, as we shall see; the Cyclops reappears in another passage in the Doxa.

The second passage that serves as an introductory signpost for the importance of the Doxa once again uses Homeric associations, this time to invoke the famous chariot race in the *Iliad* (23.262–52) where the charioteer with slower horses uses mētis to "ride on past" (παρελάσσῃ) his superior opponent and win the race.[16] As Peter Kingsley has demonstrated, this line not only invokes mētis as the means by which the chariot race is won, through which the less powerful rider uses cunning to best his opponent, but it also contains clever word play on οὔτίς / μή τίς ("no one") that echoes that of Odysseus when he famously fools the Cyclops by telling him his name is "no one": one of the most prominent feats of mētis in epic narrative.[17] This introductory signpost is especially significant because Parmenides's teaching can be understood as a means of awakening this latent ability of cunning resourcefulness that the ancient Greeks called mētis: the greatest hero of epic, Odysseus, is renowned not for his force but for his wiles, his persuasion and ability to adapt to changing circumstances like a skilled navigator (often with the divine help of Athena, daughter of the goddess Metis).[18] Mētis is a divine ability, and mortals that are known for their mētis are often acting with divine aid like Odysseus; in Parmenides's poem, the kouros is receiving divine wisdom from the unnamed goddess.

The Cyclops is an important thematic image because the creature with one eye sees a world without depth. Human beings, by contrast, are accused of a kind of double vision: mortals are "twin heads" who believe a confused version of reality, where "being and non-being are the same but not the same" (6.8–9). As we shall see, the ability to see and understand two things at once is crucial for understanding the relation between the doxa and alētheia, just as a background is often necessary to make visible a figure in the foreground. The Cyclops is without mētis. In fact, the Cyclops is an image of hubris, as evident when he tells Odysseus, "You are a fool, stranger, or have come from afar, seeing that you bid me either to fear or to avoid the gods. For the Cyclopes pay no heed to Zeus, who bears the aegis, nor to the blessed gods, since truly we are far better than they."[19] To look only to the Truth portion of the poem is to be like the one-eyed Cyclops, believing in a unified view while lacking the depth supplied by Doxa; it is necessary to have experience of how the truth *appears* to know alētheia.

Mortals do not see the absolute unity of alētheia, and even worse, the doxa is starkly presented by the goddess as a battleground of war. When the goddess says, "I tell you this whole arrangement as it seems to mortals so that no mortal can ride on past you in knowledge,"[20] the word translated as "arrangement" is διάκοσμον: a battle formation. As Mourelatos describes it, "by the use of διάκοσμος the meaning of 'order' in -κοσμος is inverted into 'segregation, division, cleavage, conflict.' The κοσμος of mortals is actually a battlefield."[21] Like the oxymoron of the "nightbright" moon, the doxa of mortals is an order that is also antithetical to order, a "divided order."[22] Mortal thinking is divided; mortals habitually bifurcate their perception and engage in binary thinking. The description of mortals as dikranoi ("twin heads") in Fragment 6 echoes this division; mortals are divided from themselves in the activity of their perception and thinking. Parmenides's teaching is a means of healing this division as the student becomes aware of these habits of thinking and gains the requisite experience offered by the poem. With these thematic and contextual prefaces in mind, this chapter will identify the experiential sites in the text and situate the Doxa within the whole of Parmenides's teaching to demonstrate its essential function.

Alētheia's Binding: Thumos and the Unshaken Heart of Persuasive Truth

Thumos ("spirit") appears in the poem's first line: "the mares that carry me as far as thumos can reach" (1.1). The kouros has not come to the goddess by accident, rather, by "Rightness (Themis) and Justice (Dike)" he finds himself "far from the beaten track of mortals" (1.27–28). This initial experience of heart is imperative in initiating the journey. The experience of thumos is needed, not that of mind (νόος) but an organ of *feeling*; in fact, it is the same word that Sappho uses to refer to her "heart" when she calls on her patroness, Aphrodite.[23] This role of thumos is especially significant because the teaching the goddess offers is about the routes that exist for *thinking*: it thus is significant that it was not thinking that led him to the goddess, but thumos. As I will demonstrate, Parmenides's text is not merely about "thinking" or "routes of inquiry" but emphasizes the limitations of mortal thinking and offers means of moving beyond those limits.[24]

If thumos is the first necessary experiential element required for passage into Parmenides's teaching, the next essential piece is stillness, hesychia. The understanding of Parmenides's journey as katabasis (roughly meaning a "descent" or "trip to the underworld") is here illuminating: as iatromantis, "healer-prophet," Parmenides would have been skilled in altered states of consciousness through incubatory practice.[25] The state of stillness is required for entry into alētheia because mortal thinking is habitually vacillating between binaries; as "twin heads" (δίκρανοι) there is motion back and forth between contraries: the motion of thinking. The only true route, as it turns out, is the route of Peitho, goddess of Persuasion; and Peitho is in the train of Aphrodite, queen of all hearts.[26] The kouros allows himself to be carried so his agency is a receptive agency, which is very distinct from some other familiar conceptions of agency such as models of mastery or the menos of the warrior on the battlefield. This receptive agency works through the practice of stillness, as the teaching of Parmenides in the alētheia describes a reality where there is no motion, no division, no separation: hence no time and space, no separate awareness. The practice of stillness allows the kouros to drop below the threshold of thought and experience an awareness distinct from the mortal habits of thinking.

Stillness is not merely ancillary to alētheia for Parmenides, as he offers a very strange and unique image of truth: Ἀληθείης εὐκυκλέος ἀτρεμὲς ἦτορ. Given the significance of thumos in initiating the journey, this choice of words is very telling: Ἀληθείης εὐκυκλέος ἀτρεμὲς ἦτορ is usually translated as "the unshaken heart of persuasive truth." If Parmenides were simply referring to some intellectual concept, alētheia would presumably be sufficient; instead, we are confronted with this strange image of an ἀτρεμές heart: a heart that is still, without motion, calm, unmoved. Such a heart would certainly *seem* to be dead. In the context of the poem, the kouros has been carried to a place "far beyond the beaten track of mortals" through gigantic doors that have opened and left a chasm: plausibly the underworld, abode of the goddess Persephone (who remains unnamed in the poem, as is fitting for chthonic deities). The identity of the goddess is also implied by the first words that she speaks in greeting, reassuring the kouros that no μοῖρα κακή (evil fate, a euphemism for death) has sent him to this place, but Themis (Rightness) and Dike (Justice).[27] This unusual and unique description—not simply alētheia but Ἀληθείης εὐκυκλέος ἀτρεμὲς ἦτορ (the unshaken heart of persuasive truth)—well exemplifies Parmenides's precise language, and that most scholars pass over this crucial phrase in silence and refer instead only to alētheia illustrates the traditional inattention to Parmenides's poetic skill and exactitude.

While I will not be focusing on the Truth portion of the poem in this chapter, it is necessary to identify the primary devices within it so they may be contrasted with the devices of the Doxa. The statements that the goddess makes in Truth have a precise character: they are circular. Her "arguments" are like lassos, binding wandering minds and fettering them to their ground. This aspect of Truth has caused much consternation and dismay to contemporary rationalistic readers of Parmenides who identify circular reasoning with fallacious reasoning; but circularity is an effective method as, throughout Truth, the goddess is actively binding the kouros and his thinking. It is not only the "heart of persuasive truth" and the "well-rounded sphere" of Fragment 8 that are bound in place and ἀτρεμές; the goddess is binding the thinking of the kouros as well. As Laura Gemelli Marciano argues, "The so-called line of argumentation works like a magical expedient that is meant to coerce and 'enchain' him and to forbid him to think of anything other than 'IS.' "[28]

The circular method of the goddess's speech in Truth does not contain "arguments" in any traditional sense because the goddess simply repeats—in multiple contexts—the essential teaching of "is" and the utter impossibility of nonbeing. The false appearance of argumentation may emerge because her speech seems to anticipate the ways in which mortal thinking, as διάκοσμος ("divided order" or "battle formation"), cannot help but separate or divide being from itself: birth, death, motion; each of these implies nonbeing because it introduces some separation or division into being. Parmenides's text is particularly prescient for suggesting deep philosophical questions regarding our use of language: words and speech by their very nature separate and divide being. Once something is "named," it is set apart, at least in thought, from everything else. The antidote to this mortal habit of naming is the goddess's statement at lines

8.38–39: "its name shall be everything: every single name mortals have invented, convinced they are all true."[29] This incredible statement poses a serious problem for language, because it means that every single name speaks the same: "is." For anything that can be spoken or thought, "is" can be the only referent. The reality that the goddess is attempting to show the kouros is one of absolute unity, with no division or separation whatsoever: this means that the use of discrete words must somehow be deceptive, since in the reality the goddess is demonstrating, there can be no discrete parts.

In Truth, the goddess binds the kouros by removing every possible route of escape, leaving no gaps open for mortal thinking to introduce separation. At 8.38–39, she removes all his words, making speech and thought essentially impossible. It is in *this* way that Parmenides's teaching, while about thinking, also indicates the limits of mortal thinking and points beyond them. Alētheia is not an intellectual content—in fact, much of Parmenides's poem demonstrates why it is not possible for mortal thinking to grasp alētheia—rather, it is an experience. This experience is possible only with very precise orientation; everyday mortal awareness has no chance of grasping it. As Empedocles says, after describing the wandering and helpless nature of mortal beings, "like this, there is no way that people can see or hear or consciously grasp the things I have to teach."[30] Just as the goddess points the kouros toward an experience that is far "beyond the beaten track of humans" because it involves letting go of thinking, Parmenides's text also implies that this movement "beyond" requires a skill more divine than mortal: mētis. In the fragment just noted, Empedocles explicitly invokes mētis and the limitation of mortals: "Mortal mētis can manage no more." Empedocles's student will only be able to grasp his teaching because the student has "come aside here," just as the kouros has traveled "far from the beaten track of mortals." The goddess's teaching offers the kouros access to the mētis needed to move beyond the forceful bonds of mortal thinking, and the method for freeing the student from these habits is not force but persuasion.

Throughout the Truth portion of the poem, the language of bonds, fetters, and binding is ubiquitous: truth is persuasive and binding by virtue of being true. The speech of the goddess is divine, efficacious speech that realizes reality instead of describing it.[31] As Marciano writes, "Truth persuades with ease precisely because it is expressed not through mere words but through words that realize themselves and bring about a direct experience: because these are experienced directly, they gave a compelling power of persuasion."³² The appearance of "argument" is effected through the use of various explanatory words like "for" or "because"; meanwhile, the statements of the goddess simply rely upon the premises she begins with: "is" must be and nonbeing cannot exist. Further quoting Marciano, "The supposed explanations are therefore a series of tautologies. But that is exactly what the goddess means to convey: the image and sense of a continuous whole which is complete in itself."³³ The "arguments" have a very real persuasive effect—and here it must be kept in mind that this poetic text would be read aloud, as its meter dictates—but the persuasive effect is not the result of careful argumentation. Instead, the repetition of the same insight—being *is* and nonbeing cannot exist—over and over again, in circular form, has an effect on the listener and evokes an experience of completeness, circularity without limit (and of course, any linguistic descriptions of it must inevitably fail to capture it).

Marciano has argued that the goddess of Truth is engaging in a magical act of binding, citing traditional magical practices in ancient Greece: for example, the mode of action found in curse tablets, which similarly attempt to "bind" the desired subject.³⁴ If the goddess that the kouros meets is Persephone of the underworld, then she is a deity particularly associated with this activity of binding.³⁵ It is significant, in the context of Parmenides's poem and its language, that binding is especially a kind of erotic magic: this sheds interesting light on the initiating role of thumos in the Proem, and the appellation "unshaken heart of persuasive truth." In this regard, the conspicuousness of the goddess Peitho in the poem also becomes more meaningful: she is associated with Aphrodite, especially in the context of erotic magic.³⁶ In binding the kouros, "The force of *Peitho* is deployed in its concrete form and leads directly to the 'unmoved heart of truth.' "³⁷ From this place of stillness, being bound and speechless, the kouros may then listen to the goddess's performance of mortal opinion, in the Doxa that follows, without being convinced by it: he has an antidote to those binary traps of mortal thinking, so engrained and habitual as to be invisible. From this strange perspective, so "far from the beaten track of mortals," the kouros can hear mortal opinions with awareness and mētis: particularly, he has the ability to be aware of multiple things at once. Thus when the goddess engages in perilous double speak in the Doxa, the kouros is able to perceive not only the surface, but the depth provided by alētheia. In turn, his experience of the doxa will provide the background in which alētheia may appear.

Dikranoi Mortals and the Necessity of Deception

After the goddess has delivered the "way of truth," she tells the kouros that she will "close her trustworthy thought and speech about the truth" and signals a change in tactics. She says, "From here forward, learn the opinions of mortals by listening to the deceptive arrangement of my words" (δόξας δ' ἀπὸ τοῦδε βροτείας μάνθανε κόσμον ἐμῶν ἐπέων ἀπατηλὸν ἀκούων, 8.51–52). "From here forward," her words are intended to deceive and will express the doxa of mortals. As the goddess has proclaimed, the ordering of mortal thinking is a διάκοσμος, a divided order, a battle formation. The goddess is about to demonstrate *how* this order is divided. To reach δίκρανοι ("twin-headed") mortals, she must use a device familiar in ancient Greek texts, a device that Hesiod names amphillogias (double speak).[38] Mourelatos has suggested that the goddess's speech in Doxa is a case of double speak, a kind of speech usually only practiced by divine speakers (or mortals with divine aid) and generally perilous and deceptive to unwary listeners. Perhaps the most famous example is Odysseus's trick in telling the Cyclops that his name is "No One." The goddess's use of the term "nothing" throughout the poem has a similar effect, particularly in Fragment 6 where mortals are described as "knowing nothing": the goddess has made it very clear that "nothing" does not exist, so this phrase takes on at least two meanings. On the one hand, it means that mortals do not know *anything*, and on the other hand, it creates a paradox by mockingly suggesting that mortals know "nothing" (if we take nothing as a "thing"). This sly remark is a great example of double speak since the goddess is able to make two very different claims simultaneously. Double speak is also associated with poetic speech: as the Muses warn Hesiod, "We know how to say many false things similar to genuine ones" (Hes. *The.* 27).[39] The goddess's speech is ironic, as Mourelatos argues, because she is speaking the doxa of mortals from the standpoint of a goddess who recognizes the "untrustworthy" character of their beliefs.[40] Listeners are warned to take what she says next with a grain of mētis.

The method of the goddess's speech throughout the poem is to meet mortals where they are: the whole ruse of the "routes," for instance, uses a concept familiar to human beings (i.e., roads) to move beyond it. The "routes" turn out to be a ruse because there is, as the goddess tells the kouros, only one route: the route of "is." And furthermore, the route of "is" very explicitly forbids any motion, which entirely negates the concept of a "route" since it is, after all, something that one travels.[41] This mischievous sleight of hand on the part of the goddess is exceptionally witty, as are her descriptions of mortals in Fragments 6 and 7 and her use of double speak in Doxa. She presents the kouros with the "routes of inquiry," and it appears that there are two, the route of "is" and the route of "is not": but the route of "is not" is immediately taken away, since it cannot be named or thought. For Peter Manchester, "It cannot be discerned or mapped. Like a black hole, no probe or possible signal returns any information. As quickly as it is formulated as a possibility, it completely self-destructs."[42] Similarly, in Truth, the goddess repeatedly uses the very language and concepts she has forbidden as impossible because they invoke nonbeing: generation, destruction, birth and death, and so forth. Once again, her tongue-in-cheek demonstration must be understood as irony since she is gently mocking mortal thinking; in Truth, she anticipates all the ways in which mortals might try to introduce separation into the "is" and cuts off all those options explicitly.

Rose Cherubin has suggested that "the goddess' account of *eon* on the road she recommends is dependent on the very opinions that it challenges," since the goddess uses terms like "wandering" and "extinguished": terms she has explicitly denounced.[43] The goddess's account must use mortal language because of the simple fact that mortals can only think and speak in language: the *experience* of alētheia cannot be spoken in mortal language because mortal language inevitably introduces separation and division, while the "is" must be an absolute unity. Thus the goddess's method in communicating to mortals must somehow manage to *indicate* beyond her words, to point to something without explicitly stating it: a task for which double speak is particularly suited. Mētis is required to grasp the multiple meanings invoked by double speak, and those who attempt to understand perilous double speak without mētis find themselves in terrible danger: the misinterpretations of oracular pronouncements are an excellent example of this in the ancient Greek tradition.[44] In Doxa, the goddess once again takes up mortal opinions, but this time she appears to be presenting them, for the most part, without criticism: but this is only to hear what she says on the surface and to miss the other dimensions of her speech produced by her skillful use of double speak.

It seems fairly uncontroversial to point out that the description of mortal perception in Fragments 6–7 (often called "the third way") parallels the opinions of mortals seen in Doxa. As Colin Smith has argued, "We can say with certainty that the barred path in Fragment 6 is the mortal path."[45] These two places in the text where mortal opinion is explicitly invoked, Fragments 6–7 (the "third way") and the Doxa, are especially replete with witty double speak. The description of mortals in the "third way" emphasizes not only the futile outcomes of binary thinking but further identifies the absolute helplessness of mortals in this state: their chests are filled with ἀμηχανίη (typically translated as "helplessness"). This concept does not mean simply "helplessness" but has connotations of an animal paralyzed by a trap, absolutely helpless and unable to make any move whatsoever: in a state of amēchania, one has no access to mētis.[46] Odysseus, when trapped inside the cave with the Cyclops, appears to be in a state of extreme amēchania, but he is able to escape this seemingly hopeless situation through his use of mētis. In Parmenides's text, habitual mortal thinking is identified with this futile helplessness and it is thumos, not noos, that allows the kouros to travel to the abode of the goddess and thus be liberated from the paralysis of mortal thinking. To begin his journey, the kouros must first experience this helplessness: and this is in fact the case for many petitioners who visit the temples of Asclepius for healing; they have tried ordinary means and recognize that their best option is simply to lie down and await divine help, accepting the limits of their own agency.[47] Only from this recognition of helplessness can the kouros allow himself to be carried and bound to persuasive truth by the goddess in alētheia.

If Truth is characterized by its persuasive qualities, mortal thinking has an opposite character: habit is violent and forceful, dragging mortals along without their consent or agency, just as their understanding is a διάκοσμος, a battle formation. At the end of the "third-way" portion of the poem, the goddess says this: "Nor let much-experienced habit force you along this way" (μηδέ σ' ἔθος πολύπειρον ὁδὸν κατὰ τήνδε βιάσθω, 7.3). The description of habit adheres to the oxymoronic formula that the goddess has been repeating throughout the poem: habit creates action that is done without awareness, so the phrase "much-experienced habit" is a paradoxical joke. The goddess has just referred to these mortals of habit as "deaf and blind at the same time"; in other words, they are certainly not in a state receptive to intelligible experience. Like the mortals in Heraclitus's estimation, they "do not recognize their own experience" (DK 17) and are "oblivious of what they do awake just as what they do asleep escapes them" (DK 1). These mortals are like the many leaves that all look the same and fall not by their own agency but because they are blown by the wind; the goddess invokes this image through the strategic use of a phrase that plays on the Homeric formula ἄκριτα φῦλα (often translated as something akin to "undiscerning crowds.")[48] Like the mortals who are carried up and away helplessly like smoke in Empedocles, their helplessness is absolute, and their thinking is entirely unable to grasp reality; as Empedocles says, they are "totally persuaded by whatever each of them happened to bump into while being driven one way, another way, all over the place. And they claim in vain that they have found the whole" (B2, trans. Kingsley 2020 [2003]). Similarly, the mortals trapped in Parmenides's "third way" hopelessly shuffle back and forth at an imagined crossroads, unable to set foot on the path of "is" because they believe they can somehow mix being and not-being; they cannot see the ways in which their habits of thinking—especially binary thinking—obscure the unity of "is" by conjuring up its so-called opposite: "is not." This hopeless shuffle is named in Fragment 6 as παλίντροπός, as turning backward on itself: a movement that goes nowhere. Fragment 6–7's grim description of mortal perception is, since the subject of the goddess's teaching is *thinking*, an image of mortal thinking from the perspective of the goddess.

The marked contrast between persuasion (πείθω) and force (βιά) is a frequent one in archaic Greek texts; perhaps the most prominent example is the contest in the *Homeric Hymn to Hermes* between the brothers Hermes and Apollo, who represent persuasion and force, respectively. In the *Iliad* and the *Odyssey*, Achilles is the greatest of warriors for his force on the battlefield, but it seems evident that Odysseus is the most admired of Greek heroes for his persuasive, crafty skills, as his famous epithet polymētis ("he of many wiles") indicates. The goddess of Parmenides's poem does not abduct the kouros, as Hades abducts Persephone with his famous mares, but the mares carry the kouros willingly: it was his own thumos that led him to be carried in this way. Thus the agency of the kouros is not a forceful agency, but a receptive agency that allows itself to be persuaded. The alternative to persuasion is to return to the habitual state of mortal perception, where human beings are dragged aimlessly by habit and without agency as "helplessness in their chests is what steers their wandering minds": hardly a tenable option, once the violence of habit is revealed. This contest between persuasion and force is played out in Parmenides's text, where Truth demonstrates persuasion while the hapless mortals of the "third way" and Doxa are subject to the violence of force. To be freed from this force of habit, the kouros is invited to hear mortal opinions from the mouth of the goddess, and this performance alters his familiar understanding of these mortal opinions, producing an ironic display of their flaws.

The Doxa begins with double speak in the wonderful syntactic ambiguity of line 8.53: Μορφὰς γὰρ κατέθεντο δύο γνώμας ὀνομάζειν, which Mourelatos translates literally as "perceptible forms, accordingly, they laid down two notions to name," pointing out that it simultaneously seems to say "they were of two minds, they vacillated."[49] In this passage, the action of "laying down names" expresses the habits of human beings as dikranoi: perception is divided into two. The ambiguity between "two notions" and "two minds" demonstrates how mortal thinking is divided: a diakosmos or battle formation. The image of a battle formation is echoed in the description of the two notions, named Night and Light, as their qualities are in vivid contrast to one another; they are opposite in character. Only at 8.54 does the goddess offer explicit criticism of these mortal beliefs, where she says that, in naming two forms, "it is not right that they should name one, wherein they have gone astray (πεπλανημένοι)." This passage echoes the description of "wandering minds" in Fragment 6; wandering has negative and even tragic connotations of being "led astray."[50] Here we might return to the theme of the moon in Parmenides's text, as this chapter began with exploration of Fragment 14's dense example of double speak and oxymoron. In Fragment 10, the moon has 'wandering works' in contrast to the 'destructive works' of the sun.[51]

Consider the relationship between the sun and moon suggested in Fragment 14 (where the moon wanders about the earth, shining with borrowed light) as an image of the relation between alētheia and doxa, respectively: if the moon has no light of its own, then how can it be shining bright? This riddle is precisely the riddle of the "acceptable" or "believable" character of the doxa: if it is not worthy of true trust, then why must it be learned?[52] In what way can mortal opinions possibly be "acceptable?" Mortal thinking is "wandering" because it is constantly in motion: oscillating between opposites (δίκρανοι), making distinctions and separating things out from other things by naming them: a theme raised in Fragment 1 and repeated once again in Fragment 19 of Doxa. The goddess says, "Thus according to belief, these things were born and now are and hereafter, having grown from this, they will come to an end. And for each of these did men establish a distinctive name."[53] While binary thinking is a primary obstacle for mortals in attempting to understand the "is," the fact that mortals use language, which is consistently conditioned by their experience of space and time, is perhaps an even greater problem: how to communicate an experience that is utterly motionless and admits of no separation when the only tools available for communication inevitably introduce motion and separation? This is precisely the situation the goddess is in when she must communicate the "is" to the kouros.

The dense and ambiguous Fragment 4 addresses this predicament by demonstrating to the kouros that time and space are—while very real since embodied mortals experience them—from her perspective, illusory or deceptive. This demonstration is key to understanding the necessity of deception and helps to shed light on the problem of *how* the doxa is "acceptable" despite the fact that it is "untrustworthy." In addition, this passage of Parmenides's text is particularly significant in demonstrating the manner in which his teaching, while about mortal thinking, points beyond the limits of mortal thinking. Fragment 4 says:

> See how it is that things far away are firmly present to your mind.
> For you shall not manage to cut off what-is from holding fast to what-is
> for it neither disperses itself in every way everywhere in order,
> nor gathers itself together.[54]

The force of this demonstration is startling: the goddess has just *shown* the kouros (as she tells him: "see how it is," suggesting he look to his own experience for verification) that space and time, from the perspective of awareness, is illusory. A "thing far away" may be absent physically while "firmly present" to noos. "Things far away" is deliberately ambiguous because it may refer to something distant in either space or in time. This demonstration can show the kouros how something he believes to be separate (as "far away" suggests) may be simultaneously, and paradoxically, present to his awareness. The appearance of separation, as the goddess has suggested in her mentions of mortal naming, is created at least in large part by the fact that mortal thinking is dependent upon, and completely immersed in, language.[55]

For the kouros to learn "all things," he must experience both the teaching of alētheia and approach the opinions of mortals from an awareness much different from that of ordinary, habitual mortal thinking. While Truth provides the requisite experience needed to experience the goddess's performance of the Doxa, the experience of the Doxa in turn illuminates the meaning of Truth. Because mortals oscillate between opposites and fail to understand their relation, their thinking has a flat or linear character; the goddess's teaching in Truth instead confronts the kouros not only with an image of the "well-rounded ball" but with a different kind of awareness, one akin to depth perception in contrast to the flat mode of mortals. This awareness cannot simply arise for the kouros without experience of the goddess's demonstrations, just as a magician must first perform the magic trick for the novice so the novice can *experience* the deception before she gleans the devices that produce the trick. In this way, Parmenides's teaching is not simply a treatise or philosophical argument about the "is" but offers instead a kind of gymnasium for mortal thinking to practice its movements and begin to imagine ways beyond them.

Notes

1. For detailed discussion of reading Presocratic texts in their poetic and religious contexts, see Struck 2004 and his analysis of Aristotle's influence on reading texts for clarity while dismissing "enigma"; see Marciano 2008 on the crucial significance of the Proem and some reasons why it has, until recently, been largely ignored in traditional scholarship. See also Detienne 1967 for the religious contexts of early ancient Greek poetic texts.

2. "[The poem] contains numerous thematic and stylistic echoes of the *Odyssey*, which are unquestionably deliberate," Gallop 1984, 4. Peter Kingsley (2020 [2003]) emphasizes this aspect of both Parmenides's and Empedocles's texts; Mourelatos 2008 (esp. chapter 1, "Epic Form") is another excellent resource for tracking these Homeric allusions in Parmenides.

3. As Eric Havelock has observed (1958, 136), "the philosopher's audience knew the *Iliad* and *Odyssey* by heart. The associations conveyed by these lines would be instinctive and automatic."

4. The long controversy concerning Fragment 7 and the anachronistic translation of "judge by reason" has caused much of the rationalistic bias that obscures Parmenides's teaching; see Kingsley 2020 [2003], 556–67 on how logos cannot possibly carry the meaning of "reason" here, but instead refers to the goddess's spoken discourse; see also Jenny Bryan's chapter in this volume.

5. See Mourelatos 2008 for analysis of double speak, especially in Doxa.

6. Translation from Kingsley 2020 [2003]: Χρεὼ δέ σε πάντα πυθέσθαι ἠμὲν Ἀληθείης εὐκυκλέος ἀτρεμὲς ἦτορ ἠδὲ βροτῶν δόξας, ταῖς οὐκ ἔνι πίστις ἀληθής. Ἀλλ' ἔμπης καὶ ταῦτα μαθήσεαι, ὡς τὰ δοκοῦντα χρῆν δοκίμως εἶναι διὰ παντὸς πάντα περῶντα.

7. Τόν σοι ἐγὼ διάκοσμον ἐοικότα πάντα φατίζω ὡς οὐ μή ποτέ τίς σε βροτῶν γνώμη παρελάσσῃ. The Greek text in both of these passages is discussed in the chapter.

8. Ἀλλ' ἔμπης καὶ ταῦτα μαθήσεαι, ὡς τὰ δοκοῦντα χρῆν δοκίμως εἶναι διὰ παντὸς πάντα περῶντα.

9. See Miller 2006 for a brilliant reading of the Proem as preparation for learning to think outside binary habits of thinking: to see "both/and" rather than "either/or." Here I note that this reading is not to be understood as a demotion of alētheia but instead as a necessary caveat: alētheia cannot be understood without the context set forth by the doxa.

10. See *Iliad* 23.272–595 for the chariot race and its context; mētis and cognate terms are consistent in the counsel Antilochus receives from his father before the race: 312, 313, 316, 318; see Kingsley 2020 [2003], 221 and 578.

11. Heraclitus and Parmenides are both concerned with the relation between opposites as well as the mortal inability to perceive reality properly; for example, compare Parmenides's Fragments 6–7 with Heraclitus DK 1, DK 17, DK 107 for similar descriptions of mortal perception.

12. The binary aspect of mortal perception will be discussed throughout this chapter but especially in the first section; mortal perception as conditioned by space and time is explicitly addressed in the evaluation of Fragment 4 in that section.

13. Mourelatos 2008 [1970], 224.

14. Mourelatos 2008 [1970], 225.

15. Mourelatos 2008 [1970], 225.

16. Mourelatos (2008) notices this connection to the chariot race but suggests that Parmenides uses this term "for metrical reasons" instead of parelthe, which in the *Odyssey* refers to the "bypassing involved in deception," citing *Od.* 13.291 where Odysseus is speaking to disguised Athena and does not recognize her; see Kingsley 1999, 221–30 for discussion of this Homeric reference to the chariot race at *Iliad* 23 and the significance of mētis for Parmenides's teaching.

17. See Kingsley 2020 [2003], 221, 578–79.

18. Vernant and Detienne (1991) have outlined the extraordinary significance of mētis in ancient Greek culture. Havelock (1958) has pointed out the multiple parallels between Odysseus's heroic travels and the language of paths and journeying in Parmenides's text; see also Mourelatos 2008 [1970], 17–25 for discussion of these themes and motifs.

19. Hom.*Od*.9.273–76.

20. Τόν σοι ἐγὼ διάκοσμον ἐοικότα πάντα φατίζω ὡς οὐ μή ποτέ τίς σε βροτῶν γνώμη παρελάσσῃ.

21. Mourelatos 2008 [1970], 231.

22. Throughout this chapter, I will emphasize the multiple examples of oxymoron in the goddess's speech: these self-conflicting terms reflect the divided and self-defeating character of mortal perception that dikranoi and diakosmos indicate. See note 41 below.

23. For thumos in Sappho, see Carson 2003 for commentary on Fragment 1, "Hymn to Aphrodite"; for discussion of thumos in ancient contexts, see Onians 1951 and Dodds 1951.

24. For a similar teaching in Eastern thought, the Heart Sutra in Buddhist tradition is an interesting parallel to this teaching of a "knowing beyond knowing": human thinking is only one kind of human awareness; thumos, for example, is another mode of "knowing" with very different qualities. While mortal thinking is structured according to language and "logic" (for example, the law of noncontradiction, etc.), other kinds of awareness are available to human beings such as the longing or desire of thumos, or the different state of consciousness produced by the practice of incubation, which is a state *between* sleeping and waking; this state was of crucial importance in ancient Greek culture and religious practice and, as Kingsley 1999 has demonstrated, Parmenides was associated with this tradition of incubation and the healing practices of Asclepius, son of Apollo.

25. See Kingsley 1999.

26. See Marciano 2008, 14–15 for discussion of Peitho and her function in erotic magic.

27. Marciano (2008, 13) points out (following Kingsley) that this phrase "in the language of epic means simply death" and serves as an "unambiguous sign that Parmenides finds himself in front of the goddess of the underworld."

28. Marciano 2008, 15.

29. τῷ πάντ' ὄνομ' ἔσται ὅσσα βροτοὶ κατέθεντο πεποιθότες εἶναι ἀληθῆ, trans. Kingsley 1999.

30. Empedocles quotes throughout are from fragment DK 2, Kingsley 2020 [2003] trans.

31. See Detienne 1996: 74, "speech endowed with efficacy is indissociable from its realization."

32. Marciano 2008, 14.

33. Marciano 2008, 19.

34. Marciano 2008, 20. See Winkler 1990 for detailed discussion of curse tablets and binding in erotic magic, especially chapter 3: "Erotic Magical Spells."

35. Marciano 2008, 20.

36. Marciano 2008, 14: "*Peitho* is a cult goddess who often belongs in the train of Aphrodite in archaic literature and iconography; the influence of *Peitho* is directly linked with magic spells."

37. Marciano 2008, 20.

38. See Elbert Decker 2021 for discussion of double speak as a common device in the Homeric Hymns, Hesiod, and the Presocratic texts of Heraclitus, Parmenides, and Empedocles.

39. Mourelatos 2008 [1970], 227; he cites Hesiod's *Theogony* 27, and 228–29 for "Quarrels, and Lies, and Tales, and Double Speak."

40. Mourelatos (2008 [1970], 222–25) emphasizes the ambiguity and irony of the Doxa, using tools from literary analysis. See also Elbert Decker 2021.

41. Paradoxical, even oxymoronic, images appear throughout the goddess's speech, such as helplessness "steering" wandering minds or the description of mortals as "knowing nothing" (both in Fragment 6); see also the discussion of

"much-experienced habit" (Fragment 7) below. Additionally, see Miller 2006 for discussion of the semantic ambiguity of achanes in the proem as both "chasm" and as the impossible object "closed chasm" (reading the *a-* as alpha privative). Miller argues that this impossible object is a kind of encounter with trying to think "nothing," a primary theme of Parmenides's teaching.

42. Manchester 2005, 112.

43. Cherubin 2017, 255.

44. *Oedipus Rex* contains some of the most tragic and/or amusing examples of misunderstood double speak, as Oedipus repeatedly misinterprets oracular statements, and even more tragically, repeatedly engages in double speak himself without knowledge that he is doing so: for example, when he says, "For it will not be on behalf of a distant friend, but for my own sake, that I shall drive away this pollution; whoever killed him may well wish to turn the same violence against me," 137–40; or "but now, since I chance to hold the power which once he held . . . I shall fight for him as though he had been my father," 258–64. These examples underscore the contrast between mētis and hubris: just like Odysseus defeating the Cyclops then bragging about it, Oedipus solves the Sphinx's riddles and then arrogantly assumes his skill in interpreting oracles.

45. Smith 2020, 285. Smith's article takes up the question of how many "routes" there are in Parmenides's text, arguing that a "three-route" structure does not easily reconcile with a teaching so steeped in binaries. While this is certainly true, I would argue that the number of paths, indeed the idea of a "path" at all, is something of a joke in Parmenides since there is, in the end, only one path ("is"). The path of "is not" is impossible and the so-called third way is simply mortals failing to recognize that they are in fact on the path of "is" since there is nowhere else to be. The three-route structure is significant, though, since ancient Greek crossroads—called trihodos, "three ways"—are in the shape of a "Y" just as the proem to Parmenides's poem brings the kouros to the krisis faced by all dikranoi mortals, even if that krisis has, as the goddess says, "already been decided" (8.16). See Kingsley 2020 [2003], 99–103 for discussion of crossroads as dikranoi and places famous for the deceptions of thieves, tricksters, sorcerers, and charlatans ("the path mortals fabricate," making ourselves the deceivers, in this case).

46. See Detienne and Vernant 1991, esp. chapter 2, "The Fox and Octopus," about traps and cunning. Amēchania is literally to have no mēchos, that is, no device or means of escape.

47. See Kingsley 1999 for Parmenides as Iatromantis, priest of Apollo.

48. See Kingsley 2020 [2003], 96 for discussion of the wordplay between akrita phula (indistinguishable crowds) and the Homeric phrase akritophullon, "indistinguishable leaves."

49. Mourelatos 2008 [1970], 228.

50. Sappho (16.9) uses this word to describe the manner in which Helen was "led astray," a catastrophic wandering, as it initiated the motions of the Trojan War.

51. Fragment 10: Εἴσῃ δ' αἰθερίαν τε φύσιν τά τ' ἐν αἰθέρι πάντα σήματα καὶ καθαρᾶς εὐαγέος ἠελίοιο λαμπάδος ἔργ' ἀίδηλα καὶ ὁππόθεν ἐξεγένοντο, ἔργα τε κύκλωπος πεύσῃ περίφοιτα σελήνης καὶ φύσιν, εἰδήσεις δὲ καὶ οὐρανὸν ἀμφὶς ἔχοντα ἔνθεν ἔφυ τε καὶ ὥς μιν ἄγους' ἐπέδησεν Ἀνάγκη πείρατ' ἔχειν ἄστρων.

52. Fr. 1.30–32: ἠδὲ βροτῶν δόξας, ταῖς οὐκ ἔνι πίστις ἀληθής. Ἀλλ' ἔμπης καὶ ταῦτα μαθήσεαι, ὡς τὰ δοκοῦντα χρῆν δοκίμως εἶναι διὰ παντὸς πάντα περῶντα.

53. Gallop's translation. Fragment 19: Οὕτω τοι κατὰ δόξαν ἔφυ τάδε καί νυν ἔασι καὶ μετέπειτ' ἀπὸ τοῦδε τελευτήσουσι τραφέντα· τοῖς δ' ὄνομ' ἄνθρωποι κατέθεντ' ἐπίσημον ἑκάστῳ.

54. This translation borrows from those of Kingsley and Gallop. Fragment 4: Λεῦσσε δ' ὅμως ἀπεόντα νόῳ παρεόντα βεβαίως· οὐ γὰρ ἀποτμήξει τὸ ἐὸν τοῦ ἐόντος ἔχεσθαι οὔτε σκιδνάμενον πάντῃ πάντως κατὰ κόσμον οὔτε συνιστάμενον.

55. This analysis offers a response to Cherubin's observation (2017, 260): "The possibility remains that the goddess might imply that we use noos to grasp correctly what makes noos possible. If this were her implication, it would pose serious problems, for it would rest on a circularity or on a begging of a question." Circularity is precisely the method and form of the goddess's teaching but need not only be understood in the modern logical terms of fallacious reasoning: Parmenides's teaching *does* demonstrate the manner in which mortals might use thinking to understand the ground of their own thinking, but with the caveat that thumos is needed to carry the kouros beyond the familiar habitual mindset of humans and mētis is required to attend with awareness to the various devices and tricks of the goddess throughout the text.

Chapter 12

Fragment 18 Revisited

Joseph B. Zehner Jr.

Femina virque simul Veneris cum germina miscent,
venis informans diverso ex sanguine virtus
temperiem servans bene condita corpora fingit.
nam si virtutes permixto semine pugnent
nec faciant unam permixto in corpore, dirae[1]
nascentem gemino vexabunt semine sexum.

[When a woman and a man mix the seeds of Venus together
from opposite blood in their veins, a power of formation,
if it maintains the mixture, fashions well-constructed bodies.
For if the powers should fight in the composite seed
and if they should not make one in the composite body,
   they dreadfully
shall harass the developing sex with a twofold seed.]
   (Fragment 18)[2]

Parmenides is the earliest Greek philosopher to show an explicit interest in the subject of embryology, according to surviving fragments. Yet, even if no earlier Presocratic fragments express such theories, there is a precedent for interest in the topic. We know, for instance, that Anaximander used embryology as a microcosmic analogy, and it is reasonable to suspect that Thales did, too.³ In this chapter, it will be suggested that the same analogy is also found in the epic poet, Hesiod, equally relevant to Parmenides. I will argue that Fragment 18 is a continuation of this tradition since embryology informs the cosmology described in Fragments 9 and 12. Furthermore, the analogy persists in Empedocles, who continues in Parmenides's footsteps.⁴

Before we can use Parmenides's contemporaries and predecessors, the fragment's original context in the fifth-century CE medical writer Caelius Aurelianus takes priority. Therefore, this essay begins with an analysis of Caelius's translation of Parmenides, comparing it to Hermann Diels's reconstruction and building on the work of more recent scholarship.⁵ The focus will be on the two terms that have caused the most interpretive issues. The first is sexus, which has no direct Greek equivalent.⁶ Therefore, this is the most likely spot in which Caelius has changed Parmenides's message. The second, dirae, has been interpreted in two different ways, sometimes as a noun signifying the Dirae, translated as "furies," and other times as an adjective, often translated as "terrible," or "dreadful." Although most scholars opt for the latter translation, the former is worthy of more consideration. That is, perhaps Parmenides described the furies harassing the developing child. If Dirae, "furies," is faithful to the original Greek fragment, then Caelius may have translated a form of the word Ἐρινύς, which suggests many exciting possible parallels, as for instance in Hesiod.⁷ Notably, Diels (1897, 116) dispenses with dirae entirely in his reconstruction, suspecting that Caelius inserted the word to support his argument and betray his own superstitions.

After an analysis of the relevant Latin passages, this essay will reevaluate the role Fragment 18 plays in Parmenides's poem. As previous scholars argued, the fragment gives insight into Parmenides's cosmology, which describes the mixture of male and female principles to create the cosmos.⁸ Furthermore, if the "furies" reading of dirae is correct, it shows that Parmenides continued to use mythological imagery and personification outside of his Proem (cf. Schrijvers 1985, 45 and Northrup 1980). There is even an opportunity to reflect upon this aspect in Truth, in particular the personification of ἀνάγκη (necessity) in bonds (ἐν δεσμοῖσιν, 8.30–31).⁹

After reevaluating the role Fragment 18 plays in Parmenides, this essay compares it with similar passages from his predecessor, Hesiod, and his contemporaries, Alcmaeon and Heraclitus. Even if we cannot reconstruct everything about the original message of the fragment, it is still representative of larger trends in the history of early Greek philosophy. The use of biological analogies is persistent, but so is the use of mythological and genealogical motifs, such as catalogues of personifications borrowed from Hesiod's *Theogony*.[10] Since we know that Parmenides used Hesiod as a model for other parts of his poem,[11] the Dirae of Fragment 18 could also make a connection to Hesiod. Based on passages from both the *Theogony* (185) and the *Works and Days* (802–4) that feature the Erinyes, this essay argues that Hesiod was the model for the Parmenides's fragment.

Of the nineteen surviving fragments of Parmenides not labeled doubtful or spurious by Diels, Fragment 18 is the only one in Latin. Translated by Caelius, it describes two types of embryonic processes, which both begin with a mixture of the parents' "seeds." One process is called "harmonious," since the seeds mix to become one, resulting in a "well-constructed" body. The other is characterized by strife: the two parents' seeds fight, never fully become one, and, as a result, "harass" (vexabunt) the sex of the child.

Based on the Latin version of the fragment alone, it is not entirely clear what Parmenides originally meant to describe. It appears in a chapter titled, "On effeminate or submissive men whom the Greeks call malthacos" (De mollibus sive subactis, quos Graeci malthacos vocant) in the work *On Chronic Diseases* (4.9). Caelius's work is, in turn, dependent on that of his Greek predecessor, the second-century physician Soranus, who was also Caelius's source for the fragment.[12] It is fair to assume that Soranus included the fragment for the same reason that Caelius did: to give an example of an ancient theory about the origins of homosexuality as an affliction or disease.

Since the fragment on its own does not explicitly mention homosexuality, one must rely on Caelius for this interpretation, but it remains possible that Caelius, and Soranus before him, misrepresented Parmenides's views. Furthermore, since the original words of Parmenides are lost, we can question the accuracy of Caelius's translation: he may have taken liberties, changing Parmenides's text to suit his own purposes. Did Parmenides give a theory about the origins of homosexuality or was his original message about something else?

Many scholars take Caelius at his word and believe that Parmenides published an embryological theory about homosexuality in his poem.[13] Others have accused Caelius of modifying the fragment, arguing that Parmenides must have intended something else.[14] Of the latter group, the theories about the original message vary, but among the most popular is that Parmenides intended to explain the origin of intersex individuals, whose sexual characteristics might not conform to what are considered typically male or female bodies.[15]

There is yet more controversy attached to the apparently negative language with which the fragment characterizes the strife-ridden embryos. Some scholars suspect that Parmenides meant to condemn or marginalize one group of embryos over the other. They, in turn, implicate Parmenides, and even Greek philosophy as a whole, in biases toward certain genders or sexes.[16] In a 2019 article, Rose Cherubin has responded to the implication of gender bias in Parmenides. Regarding Fragment 18, Cherubin (2019, 37) initially points out that both mother and father contribute something to reproduction, contrary to some Greek views. Moreover, taken with Fragments 12 and 17, it seems Parmenides believed that at least some mixtures of male and female principles are favorable (Cherubin 2019, 38), and even that femaleness was associated with other positive qualities (e.g., light and heat, as opposed to dark and cold), inverting some traditional associations (Cherubin 2019, 36; cf. Journée 2012, 296–301 and 306–8). In other words, it is not a bad thing that a female principle is a part of us all. Furthermore, Cherubin describes many aspects of Caelius's and Soranus's discussion of Fragment 18 that probably did not belong to Parmenides's poem. For instance, Caelius takes the fragment to describe males, while the fragment itself seems to give equal attention to both male and female principles, indicating that Parmenides did not refer to "phenotypical males" exclusively. In addition, the use of derogatory terms by which Caelius and Soranus describe some sexual behavior, namely molles, subactos, and malthacos, are from a time later than Parmenides and, therefore, were not used in his poem (Cherubin 2019, 38–39). Finally, Parmenides does not blame the "clashing-seeded" people. Even if their bodies are not "well-constructed," this characteristic does not make them "intrinsically inferior" or "morally suspect" (Cherubin 2019, 40–41). In the end, it remains possible that Parmenides challenged patriarchal norms instead of supporting them.

Through analysis and comparison of both the fragment and its context, Cherubin convincingly shows that the biases belong to Caelius and Soranus and not to Parmenides. Aside from some significant differences in detail, a similar point was made by Diels (1897, 114–16, cf. Cherubin 2019, 39–40). There is no evidence for attributing Caelius's homophobia to Parmenides and it may, in fact, be anachronistic.[17]

This brief summary of previous scholarship indicates what is at stake in discussions of the fragment, but, if Caelius is wrong, it also shows that we still do not know for certain what Parmenides's original message was.

As a result, we still primarily rely upon Diels's critical work reconstructing the Greek. Although it has flaws recognized by all, no other reconstruction exists notwithstanding the many worthy suggestions made by scholars about particular words or phrases.

There are, however, limits to what we can hope to accomplish here. As one of very few short fragments from the Doxa, there simply is not enough surviving text to reconstruct Fragment 18 in any demonstrable fashion, a point that Diels no doubt understood despite his pioneering effort. Furthermore, the fragmentary status of Parmenides and the other Presocratics is not the only problem. There is also the rarity of the subject matter, since Parmenides's Fragments 17 and 18 could in fact be the earliest extant embryological theories from all Greek literature.[18]

Although there are similar theories, it is entirely uncertain whether the details lost from Parmenides's poem would agree with other writers. For instance, like Parmenides, Alcmaeon was a "two-seed" theorist, claiming both the man and woman contribute semen (A13 DK). We also know that Alcmaeon believed the sex of the child was determined by whichever parent supplied a greater amount of semen: an example of so-called epikrateia theory (A14 DK). It is, however, not clear whether Parmenides believed the same thing about sex determination, and, furthermore, none of Alcmaeon's fragments or testimonia addresses the two different types of embryos that we see in Fragment 18.[19] The same sort of uncertainty will apply to other theories from later writers.

With a lack of parallels, it is especially difficult to infer what types of individuals the strife-ridden embryos are meant to describe. As a result, our reliance on Caelius and other intervening sources may increase, but there is always a danger in projecting the details from later authors onto Parmenides. It is, however, necessary that before we can confidently understand the fragment's role in Parmenides's poem and its relationship to the rest of early Greek philosophy, Diels's reconstruction and the other suggestions must be reassessed on the basis of *all* available evidence. Some of this evidence ought to include earlier writers: influential philosophers and poets who are ignored more often than the later medical writers.

Fragment 18 and Caelius Aurelianus

Caelius produced Latin editions of three different Greek works: *On Diseases of Women, On Acute Diseases,* and *On Chronic Diseases.* While Soranus's gynecological treatise survives, the other two Greek works are lost. Since both versions of *On Diseases of Women* survive, one might hope to infer how closely Caelius's other two translations reflect the two lost works of Soranus, except that, as Philip J. van der Eijk (2015, 305) puts it, "The comparison is complicated by the different status of the extant works; for Soranus's *Gynaecia*, his only major surviving work, is a very specialized treatise, and a very practical one at that, whereas Caelius's *Acute* and *Chronic Affections* deal with what may be called the central concern of Methodism, namely the diagnosis and treatment of diseases."[20]

Based on van der Eijk's comment, it seems best to infer Caelius's dependence on Soranus from the text of *On Chronic Diseases* itself. For the purposes of this chapter, however, the inference is only relevant insofar as it might help isolate what embellishments Caelius or Soranus may have imposed upon Parmenides. The focus, therefore, will be on the places in which Caelius betrays his own biases, as well as where the quoted fragment seems to be at odds with the surrounding context, indicating that, despite his biases, some details in Caelius remain faithful to Parmenides.

Since Caelius was active nearly a millennium after Parmenides wrote his poem, it is doubtful that he had access to the entire work and more probable that he was only familiar with the few lines presented here. The often-quoted passage by Simplicius on the rarity of Parmenides's poem is relevant (A21 DK), even if Caelius was writing up to a century earlier. This suspicion is also confirmed by when Caelius introduces Fragment 18 (*Tardes Passiones* 4.9.134):

> Parmenides libris quos *De natura* scripsit eventu inquit conceptionis molles aliquando seu subactos homines generari. Cuius quia graecum est epigramma, et hoc versibus intimabo. Latinos enim ut potui composui ne linguarum ratio misceretur.
>
> [In the books he wrote *On Nature*, Parmenides says that effeminate or submissive men are produced at some time in the event of conception. And since his Greek is an epigram, I will translate this in verse. For I composed Latin verses as well as I was able so that the quality of the languages would not be mixed.]

Diels remarked that the terms libris and epigramma imply that Caelius was treating the text of Soranus freely (1897, 114). Diels does not elaborate, but perhaps he saw an inconsistency in the plural libris and the singular epigramma, or he took the terms as a clue that Caelius did not know the length of Parmenides's original poem. Responding to Diels, Piet Schrijvers rejects the suspicion, claiming that the different terms could be for the sake of varatio (1985, 42). Schrijvers refers to Mario Untersteiner, who argues that the term epigramma reflects the brevity with which Parmenides treated his themes (1985, 68 n. 79; cf. Untersteiner 1958, 166). Untersteiner's point, however, assumes both that Caelius had the whole poem at his disposal and that Parmenides's interest in embryology did not play a larger role in the poem, a difficult argument to secure for a fragmentary work. It is more likely that the term epigramma, "short poem" (*OLD* s.v. epigramma 3), implies that Caelius only had the six hexameters to work with and nothing more. The term libris may be translated from the text of Soranus, but epigramma is from Caelius's own statement about what he intends to translate, referring to the six Greek hexameters of Parmenides (cuius).

The meaning behind Caelius's explanation for why he translates Parmenides has not been fully appreciated. Most scholars take it to mean that Caelius translates accurately (e.g., Schrijvers 1985, 42), but since the statement emphasizes the reason why he translates in the first place (enim), the focus should be less on his abilities and more on the purpose clause: "so that the languages would not be mixed." On the one hand, this could be a reflection of Caelius's task, to translate Greek medical texts into Latin. On the other hand, Caelius does often use Greek terminology throughout his work. Giving the Greek name for each disease, Caelius "mixes" the languages in every chapter. Avoiding the mixture of languages here implies that Caelius did not use transliterated Greek when he could have; he was Latinizing the hexameters more than was necessary. Furthermore, Caelius is making a pun on the fragment itself: he wants his text to be like the unified, "well-constructed" body, suggesting that, if he did mix Latin and Greek, the lack of unity may have negative consequences similar to what the child suffers in Parmenides. It is, therefore, unnecessary to suppose that Caelius's Latin version should follow the Greek especially closely, if he is going out of his way to avoid Grecisms such as Erinyes. In the end, Caelius's presentation expresses pride in not only his accuracy but also his creative contribution.

Some elements of Caelius's hexameters even suggest that he adapted Parmenides to Latin models. For instance, the phrase permixto semine is Lucretian (*DRN* 2.585; 2.687, cf. 6.789), referring to the mixture of atoms.[21] The parallel suggests a number of possibilities. It could be coincidental, or Caelius is using Lucretius to help him express Parmenides in Latin, which is a reasonable strategy. If he did use Lucretius, it may be a sign that Caelius departed further from the original or even a testament to Caelius's skill as a translator.

The subject-verb combination dirae ... vexabunt suggests yet another parallel from Latin poetry, but from an elegy by Propertius (2.20.27–31):[22]

> cum te tam multi peterent, tu me una petisti:
> possum ego nunc curae non meminisse tuae?
> tum me, vel tragicae, <u>vexetis, Erinyes</u>, et me
> inferno damnes, Aeace, iudicio;
> sitque inter Tityi volucres mea poena iacere,
> tumque ego Sisyphio saxa labore geram.

> [Although so many were courting you, you alone courted me:
> Am I not able now to remember your care?
> Then you tragic Erinyes may persecute me, if you wish,
> And you, Aeacus, can condemn me with an infernal judgment;
> And let my punishment be to lie among the birds of Tityus,
> And then I will carry rocks in Sisyphian toil.]

On the assumption that the term dirae in Fragment 18 translated Ἐρινύες, the parallel in Propertius suggests a number of possibilities. Perhaps Caelius, upon seeing Ἐρινύες in Parmenides, but not wanting to mix the languages, borrowed the Latin verb, vexo, while changing the subject from Erinyes to Dirae. The elegiac parallel also suggests something about the meaning of Erinyes in Latin poetry. Although they might simply be typical underworld punishers in the Propertian passage, the Erinyes also might cause the poet psychological turmoil if he should forget his loyalty to his lover. For Caelius, the psychological effects of the Erinyes on their victim ultimately support how he is using the Parmenides fragment.

The Latin parallels might also suggest that the style and word order of the original Greek could have been very different from the Latin version. In Diels's reconstruction, for instance, there are some places in which he departs from the Latin significantly (1897, 44):

> ἀλλ' ὅταν ἄρσεν' ὁμοῦ καὶ θήλεα κύματα μίσγῃ
> Κύπριδος, ἔκ τε φλεβῶν δύναμις σὺν ἐναντία πλάσσῃ,

ἢν μὲν κρῆσιν ἔχησιν, εὔκτιτα σώματα τεύχει·
ἢν δὲ δίχα φρονέωσι βροτῶν ἐν σπέρματι μεικτῷ
μηδὲ φύωσιν ὁμὴν δυνάμεις ἐνὶ σώματι μεικτῷ,
γεινομένην διφυεῖ σίνοιντό κε κύματι φύτλην.

[But whenever male and female seeds [?] of Kupris
mix,[23] and if a power from the veins molds opposites together,
and if it maintains mixture in this case, it fashions well-constructed bodies.
But if the powers in the mixed seed disagree
And they do not grow together in the mixed body,
They would harm the stock [?] with a double-natured fetus [?] as it is being born.]

Some differences between Diels's version and the Latin are minor: even if incorrect, it is best to focus on the major differences first. Why, for instance, does Diels leave out a word for "blood"? Not only do the nominative αἷμα and genitive αἵματος fit hexameters easily, but both forms appear in Empedocles (nom.: 98.5, 100.6, 105.3; gen.: 105.1 DK). Furthermore, the theory that blood is the origin of semen is distinct from other theories, such as Alcmaeon's encephalo-myelogenesis theory that semen comes from bone marrow and the brain (A13 DK), and the Hippocratic pangenesis theory that it comes from every part of the body (*On Generation* 3 [7.474 Littré], cf. Wilberding 2015 and Lesky 1950, 1301–4). Since the Latin mentions blood, the Greek should too. Perhaps Diels thought φλεβῶν, "veins," implied blood sufficiently, but the phrase ἔκ τε φλεβῶν . . . ἐναντία seems weaker than the Latin's diverso ex sanguine.

Even greater differences from the Latin occur in Diels's fourth line. The phrase δίχα φρονέωσι, for instance, takes us further away from the physical altercation that pugnent could describe. Although he does not mention this in his commentary, it appears that Diels took inspiration from a fragment of Philolaus: "For harmony is the union of diverse mixtures of things and the agreement of things that disagree" (ἔστι γὰρ ἁρμονία πολυμιγέων ἕνωσις καὶ δίχα φρονεόντων συμφρόνησις, 10 DK). There is a problem, however, since there is no precedent for the combination δίχα φρονέω in hexameter, nor is it common in prose (Philolaus is the earliest example). It is also possible that Diels was influenced by Caelius, and even Soranus, when adopting this phrase, but this requires some explanation.

Throughout the chapter "On effeminate or submissive men" (De mollibus sive subactis), Caelius explicitly refers to Soranus only once. The reference shows that both Caelius and Soranus characterize the disease in the same way: as a disease of the mind. Caelius begins the chapter with the following: "For this does not occur in human behavior because of Nature, but because lust defeated modesty and subjected parts certainly not meant for obscene uses" (Non enim hoc humanos ex natura venit in mores, sed pulso pudore libido etiam indebitas partes obscenis usibus subgavit, 4.9.131). He elaborates that the symptoms are "foreign to diseases of the body, but rather belong to a corrupt mind" (. . . quae sunt a passionibus corporis aliena, sed potius corruptae mentis). Caelius finally adds that Soranus held a similar opinion: "For it is, as Soranus says, a disease of a shallow and repugnant mind" (Est enim, ut Soranus ait, malignae ac foedisssimae mentis passio, 4.9.132).

The distinction between body and mind is one that Caelius also brings into his explication of Parmenides's Fragment 18 (4.9.135):

> vult enim seminum praeter materias esse virtutes, quae si se ita miscuerint ut eiusdem corporis faciant unam, congruam sexui generent voluntatem. Si autem permixto semine corporeo virtutes separate permanserint, utriusque veneris natos adpetentia sequatur.
>
> [For [Parmenides] wants the powers of seeds to be beyond matter, and if these would have mixed so as to make one power of the same body, they would produce a disposition that agrees with the sex. But if the powers should remain separate when the material seed is combined, then an appetite for both types of love would pursue the children.]

Caelius is establishing a dualism that does not agree with Parmenides. On the one side, we have the physical: matter, body, semen. On the other, Caelius posits a transcendent category, "beyond matter," which seems, for him, to include not only the virtues of the fragment, but is somehow also connected to the disposition (voluntas) of the child and their appetite (adpetentia).[24]

According to Caelius, the abstract powers of the fragment seem to have a direct effect on the psychology of the child. It is important to note, however, that Caelius embellishes here. For instance, in the first three lines, the fragment says that one virtus fashions a well-constructed body, describing the physical development of the child. Caelius's paraphrase adds the idea that the virtus "produces a disposition." In the second half of his paraphrase, there is a similar embellishment. If the two powers remain separate, an appetite "pursues" (sequatur) the child. The paraphrase corresponds to the last lines of the fragment, describing the dirae harassing the child. Here, Caelius does not say specifically that the virtutes produce the appetite. Although that could be what is meant, it remains possible that dirae does not refer to the virtutes, rather it refers to the Dirae, the Furies (Ἐρινύες) in pursuit, an action well-suited to their character.[25]

The presence of the furies in the original fragment could also be the source of the psychological interpretations that both Caelius and Soranus have thrust upon Parmenides. In both Latin and Greek poetry, the Erinyes have been known to have a psychological effect on their victims (Johnston 2006). For instance, in the *Aeneid*, Allecto, who is called one of the Dirae Deae (e.g., 7.324) as well as an Erinys (7.447) causes anxieties, anger, and fury (curae, irae, 7.345; furibunda, 7.347). Furthermore, Vergil describes the goddess affecting the minds (animam, 7.35), and hearts (pectora, 7.349) of her victims, Amata and Turnus (cf. Williams 1973, *ad* 323f.). The furies also occasionally do similar things in Greek poetry, as when, in Homer's *Odyssey*, an Erinys places madness (ἄτη) in the mind (φρένες) of Melampus (15.233–34); also, in Aechylus's *Eumenides*, the furies describe their own activities (328–33):

ἐπὶ δὲ τῶι τεθυμένωι τόδε μέλος, παρακοπά, παραφορὰ φρενοδαλής, ὕμνος ἐξ Ἐρινύων δέσμιος φρενῶν, ἀφόρμικτος, αὐονὰ βροτοῖς.

[For the sake of victim, the following song: Frenzy, mind-ruining derangement, a hymn from the Erinyes, binding of minds, without a lyre, withering to mortals.]

Because the Erinyes frequently cause madness, both Soranus and Caelius could draw on this aspect to support their views on homosexuality. If such an interpretation holds true for Parmenides, then Diels's phrase, δίχα φρονέωσι, makes sense; but then we must also accept the continued influence of this interpretation on even the most recent scholarship, as the recent edition by André Laks and Glenn Most labels Fragment 18 with the topic "determination of the character [sc. of the child]" (2016, 11).

It is worth mentioning, however, that in none of the surviving fragments does Parmenides describe the mind-body dualism that Caelius attributes to him. The main dualism of the Doxa is the one between night and light/fire (8.53–59; 9; 12, discussed below). Although many other pairs of opposites are said to correspond respectively to night and light, none of those pairs amounts to what Caelius is implying here.[26] Furthermore, in at least one interpretation of Fragment 3, Parmenides seems to claim that being and thinking are the same (τὸ γὰρ αὐτὸ νοεῖν ἐστίν τε καὶ εἶναι).[27] Whether this claim should apply to the cosmology of the Doxa is open to debate. Nevertheless, the burden of proof lies with those who wish to claim that Parmenides *did* discuss things that existed *beyond matter*. Even the most abstract "Being" is once described in physical terms, "like a mass of a circular sphere, equal from its middle to all sides" (εὐκύκλου σφαίρης ἐναλίγκιον ὄγκωι, μεσσόθεν ἰσοπαλὲς πάντηι 8.43–44), and the goddesses whom the kouros meets can be reached by chariot (1). Therefore, the transcendence in Caelius's explication of Fragment 18 is more easily explained by the influence of later thinkers on his reading of Parmenides than it is by Parmenides's own philosophy. After all, the emphasis of Fragment 18—even in the Latin—seems to be on physical processes and only secondarily (if at all) on psychology.

In the Latin, the verb vexabunt brings similar confusion. The *OLD* ensures the main sense of the verb is physical, for example, "to apply constant blows to . . ." (s.v. vexo 1), but it shades into more abstract and emotional connotations, for example, ". . . to distress . . ." (s.v. 5). Because of the two sides of vexabunt, translating by the English verb "vex," as A. H. Coxon does, shows preference for the psychological side of the Latin verb over the physical. By contrast, Leonardo Tarán's translation, "torment" (1965, 172), is to be preferred, even over Laks and Most's "disturb" (D49). Furthermore, Diels's choice of Greek verb, σίνομαι, is difficult to surpass, since, like vexo, it can refer to physical attacks as well as emotional impact. For instance, the verb describes the effect of shame (αἰδώς) on men in the *Iliad* (24.45), as well as Scylla's attacks on Odysseus's men in the *Odyssey* (12.114). Despite the accuracy of Diels's choice to capture the ambiguity, whether the same ambiguity applied to Parmenides's word choice remains uncertain.

The meaning of vexabunt, its subject, dirae, and its object, sexum, amount to the central issue of interpreting and reconstructing Fragment

18. Suppose, for instance, that Diels was right about sexus and Parmenides wrote φύτλη. The Greek word means "stock, generation, or race" (LSJ s.v.). It appears only twice in the Archaic period: in Pindar (*Pythian* 9.33; *Olympian* 9.55). Afterward, it is conjectured to appear in the papyri of Poseidippus (*Epigrammata* 22), and not again until the first century BCE.[28] Therefore, φύτλη is a peculiar choice of word for Diels. Crucially, it does not mean "sex," nor even "gender," although it can reasonably refer to a child of any sort. If it was Parmenides's word, then the fragment could have been about any number of afflictions, deformities, or even miscarriages. Finally, this raises the possibility that the fragment may not have been about issues of sex or gender at all.

Alternatively, instead of φύτλη, Parmenides could have written something to justify Caelius's choice of word, sexus, but the alternatives are just as ambiguous. It is more typical of Greek authors to refer to sex and gender somewhat periphrastically. For example, Hesiod calls those descended from Pandora "the race of female women" (ἐκ τῆς γὰρ γένος ἐστὶ γυναικῶν θηλυτεράων, Hes. *The.* 590). The phrase "female tribe" also occurs in Aristophanes's *Thesmophoriazusae* (τὸ γυναικεῖον φῦλον, 786). These phrases suggest that either term, γένος or φῦλον, cannot stand for "sex" on their own without help from the surrounding context.

It may in fact be that no Greek term has the same connotations as Latin sexus. In *Sexing the World*, Anthony Corbeill argues that the Romans exploit the relationship between grammatical gender (genus) and biological sex (sexus) to order their world. He writes, "By dividing their world into discrete sexual categories, Latin vocabulary works to encourage the pervasive heterosexualization of Roman culture," and adds that such a practice has "little extant precedent in Greek tradition" (2015, 9). The tendency Corbeill describes depends upon the existence of the two closely related but distinct concepts, of which one, genus, does have a Greek analogue (γένος), while the other, sexus, does not.

Without straying too far from our topic, Corbeill's argument about Latin language and culture is relevant to the reconstruction of Parmenides's Fragment 18. It emphasizes that the Latin term sexus implies a greater division between the sexes than any corresponding Greek term. The etymology of sexus, from seco, "to cut," re-enforces the point (Corbeill 2015, 8; cf. Szemerényi 1969, 977–78; de Vaan 2008, 560–61). It is therefore reasonable to conclude that when Caelius opts for sexus, he is embellishing whatever original term stood in the Greek to illustrate the affliction of homosexuality as he, and Soranus before him, saw it. Although one may object that Soranus's prior inclusion of the fragment helps to guarantee Caelius's interpretation and, in turn, may even justify the embellishment, it is still possible that Soranus excerpted the hexameters selectively for his argument and thereby contradicted Parmenides's true message.

This is not to say that Parmenides had nothing to say about the determination of the sex of the child. Fragment 17, quoted by Galen (6.48=17a1002 Kühn), describes so-called Left-Right theory: "Boys on the right, girls on the left," (δεξιτεροῖσιν μὲν κούρους, λαιοῖσι δὲ κούρας). According to the interpretations of commentators, the child's sexual characteristics could be determined by the position of the fetus in the mother's womb or which testicle the semen comes from, as mentioned in Aëtius (5.7.4=A53 DK, cf. Tarán 1965, 263–67).

It is, however, assumed too often that Fragment 17 should inform our interpretation of Fragment 18: there are conflicts between the two fragments that suggest otherwise. First, it appears from Fragment 18 that the goal of conception is a kind of unity or harmony, while Fragment 17 describes the predominance of one side over another, whereby left corresponds to the female sex and right corresponds to male. Many have tried to read the same kind of epikrateia theory into Fragment 18 (e.g., Censorinus 6.5=A54 DK; cf. Lesky 1950, 50), but it can be argued that the physical process described in the first three lines of the fragment does not include a predominance of any kind. Instead, there are two options: unity (temperiem, unam) or division (pugnent, diverso, gemino). It seems relatively obvious what the unity results in: successful birth without any complications. But what precisely is the result of the "fighting"? If I am correct in casting doubt upon the object of vexabunt, sexus, then many of the common interpretations of Fragment 18 need to be reconsidered.

Fragment 18 and Parmenides

The subject of vexabunt, namely dirae, is generally assumed to refer to the virtues two lines above. It operates as a predicate adjective, translated adverbially as "terribly" or "dreadfully." Diels leaves a corresponding Greek word out of his translation because, as he believes, Caelius added the word betraying his own superstitions about homosexuality. Nevertheless, Diels does not dispense with all of Caelius's negativity, as his verb choice, σίνοιντο, brings the same connotations as vexabunt. In any case, Diels clearly accepts virtues as the subject of vexabunt.

Interpreting virtutes, Diels suggests it translated the Greek δυνάμεις and scholars have readily adopted Diels's suggestion. This raises a question about what the δυνάμεις are in Parmenides and whether the action described by the verb vexabunt appropriately describes their role in Parmenides's cosmos.

The δυνάμεις interpretation is based on another of Parmenides's Doxa fragments (9, DK trans.):

αὐτὰρ ἐπειδὴ πάντα φάος καὶ νὺξ ὀνόμασται
καὶ τὰ κατὰ σφετέρας δυνάμεις ἐπὶ τοῖσί τε καὶ τοῖς,
πᾶν πλέον ἐστὶν ὁμοῦ φάεος καὶ νυκτὸς ἀφάντου
ἴσων ἀμφοτέρων, ἐπεὶ οὐδετέρωι μέτα μηδέν.

[But since all things have been named Light and Night
and the names are given to various things according to their
    powers,
everything is full of both light and invisible night
together equally, since nothing is without either one.]

The message of these lines is obscure, especially since Parmenides raises questions about the reliability of language in his claim that all things are named "light and night." Nevertheless, it seems reasonable to infer that, since all things are full of light and night, they are the two basic, constitutive principles of the cosmos described in Doxa.[29] If this is correct, then they appear to have corresponding, though perhaps separate, δυνάμεις that justify some things being named "Light," and others being called, "Night." Both Tarán and Coxon claim that the δυνάμεις are equivalent to the σήματα mentioned at 8.55, the beginning of Doxa. If the identification is correct, then the δυνάμεις are "signs" posited by mortals to identify the "mutually opposed" elements of the universe (cf. Coxon 2009 [1986], 233). Additionally, as "signs," the δυνάμεις of Fragment 9 seem limited to something linguistic or conventional (cf. Kirk, Raven, and Schofield 1983, 256). Otherwise, the signs might be observable phenomena in nature that are themselves open to interpretation, as the crane that gives the sign to Hesiod that it is time to plow (*WD* 450).

In either case, it is not clear whether these δυνάμεις of Fragment 9 are the virtues in Fragment 18. There, the virtues seem to be more than mere "signs" but perform physical actions: they form and build bodies, fight or unify. If they are the same, then perhaps the δυνάμεις in

Parmenides's cosmology are also more than linguistic or conventional, but real properties of things. If this interpretation is correct, then it seems that Caelius calling the virtues "beyond matter" is overstepping. As a result, it seems best to limit the reach of these powers to the physical processes, mixing opposite forms to produce things in the universe. What Caelius wishes to claim, however, is that if the powers do not mix, certain desires or appetites (adpetentia) "pursue" (sequatur) the child. Even in Caelius's paraphrase, the virtues are not the appetites themselves nor even the direct cause of them, as they could be a by-product of the lack of mixture.

Fragment 12 may provide evidence that the δυνάμεις, or virtutes, describe corresponding pairs of opposite properties (DK trans.):

αἱ γὰρ στεινότεραι πλῆντο πυρὸς ἀκρήτοιο,
αἱ δ' ἐπὶ ταῖς νυκτός, μετὰ δὲ φλογὸς ἵεται αἶσα·
ἐν δὲ μέσωι τούτων δαίμων ἣ πάντα κυβερνᾶι·
πάντων γὰρ στυγεροῖο τόκου καὶ μίξιος ἄρχει
πέμπουσ' ἄρσενι θῆλυ μιγῆν τό τ' ἐναντίον αὖτις
ἄρσεν θηλυτέρωι.

[For the narrower [sc. rings] were filled with unmixed fire,
and the ones beside them were filled with night, and among
   them a portion of flame rushes about,
and in the middle of these is a Daimon who steers all things.
For she rules over the hateful birth and mixture of all things,
by sending female to mingle with male and in turn the
   opposite, male with female.]

The first lines of the fragment describe astronomical features and the role the two principles, light and night, play on a large scale (cf. A37 DK). There is also a goddess who "steers" all things, implying that she plays a role similar to Love in Empedocles: the goddess moves things and causes them to combine, which Empedocles labels "birth" according to convention (B9 DK). The fragment continues to describe her explicitly erotic role in the cosmos, as she not only combines the principles of light and night but also sends male and female to combine with one another, resulting in what Parmenides calls "hateful birth."[30] The juxtaposition of erotic combination with cosmic and physical combination suggests that all things in the cosmos are not forms of light or night alone but additionally male and female.

The fragment describing Eros as among the first gods in Parmenides's cosmology is also relevant here (Plato, *Symposium* 178b8–10 = fr. 13 DK):

> Παρμενίδης δὲ τὴν γένεσιν λέγει πρώτιστον μὲν Ἔρωτα θεῶν μητίσατο πάντων.
>
> [Parmenides says about Genesis: she devised Eros first of all the gods.]

Although it is disputed who the subject of "devised" (μητίσατο) is, some have claimed that it is the same goddess who appears in Fragment 12; others claim the goddess is Δίκη from the Proem.[31] Whoever it is, by positing Eros as among the first gods, Parmenides is following in Hesiod's footsteps, who claimed that Eros was among the first four gods in his *Theogony* (116–22). In this case, the δυνάμεις of Fragment 9 could also describe the forces and mechanisms by which light and night, male and female, and all other corresponding pairs of opposites combine in Parmenides's cosmos according to the goddess's wishes.

The inclusion of the goddess and the god Eros in Parmenides's cosmology allows for the possibility of other gods and mythological figures to have appeared in the lost portions of the Doxa. A testimonium in Cicero supports the idea even further (*De Natura Deorum* 1.28.14–18 = A37):

> multaque eiusdem [sc. Parmenides] monstra, quippe qui bellum qui discordiam qui cupiditatem ceteraque generis eiusdem ad deum revocet . . .
>
> [And many portentous things are in Parmenides, the very one who traced War and Strife and Desire and others of this same kind back to God . . .]

Cicero mentions a few figures by name, but tantalizingly passes over an undetermined number of others. As Mark Northrup has convincing argued, the Latin terms allude to Hesiod's catalogue of Night in the *Theogony* (224–32):

> ... μετὰ τὴν δ' Ἀπάτην τέκε καὶ Φιλότητα
> Γῆράς τ' οὐλόμενον, καὶ Ἔριν τέκε καρτερόθυμον.
> αὐτὰρ Ἔρις στυγερὴ τέκε μὲν Πόνον ἀλγινόεντα
> Λήθην τε Λιμόν τε καὶ Ἄλγεα δακρυόεντα
> Ὑσμίνας τε Μάχας ...

> [and afterwards she bore Deceit and Desire,
> and destructive Old Age, and she bore stout-hearted Eris.
> But hateful Eris bore painful Toil
> and Oblivion and Famine and tearful Pains
> and Fights and Battles ...]

Cicero's words bellum, discordia, and cupiditas respectively allude to Ὑσμῖναι/Μάχαι, Ἔρις, and Φιλότης in Hesiod. Based on this connection, it can be assumed that Cicero's "others" refers to other members of the Catalogue of Night.

It is, therefore, also necessary to address Coxon's suggestion that Dirae translates a different member of Night's catalogue, κὴρ and not Ἐρινύες. For evidence, Coxon provides a parallel from Homer: Patroclus's death was assigned to him as he was being born, just as the Dirae vex the embryo as it is born in the Parmenides's fragment (Hom.*Il.* 23.78–79):

> ἀλλ' ἐμὲ μὲν κὴρ ἀμφέχανε στυγερή, ἥ περ λάχε γιγνόμενόν περ

> [But a hateful death surrounded me, which came to preside over me even as I was being born.]

Note how the participle γιγνόμενόν matches the nascentem of the Caelius's translation. Coxon, furthermore, argues that Parmenides is using the Homeric parallel to allude in turn to the tradition that Achilles and Patroclus were lovers. Coxon, therefore, assumes that Caelius accurately represents the meaning of Parmenides's hexameters. Coxon's suggestion may be plausible, but it is far from conclusive.

Would a Latin translator use Dirae for κῆρες? While it does not seem impossible, the best examples for translations of κῆρες suggest that they would not. Both Cicero and Hyginus record Latin versions of Catalogues of Night, whose ultimate poetic predecessor must be the catalogue as it stands in Hesiod, which includes the keres at *Th.* 211 and *Th.* 217 (cf. Schrijvers 1985, 45 and 68 n.84). Cicero's version reads (*De Natura Deorum* 3.44.3–15):

> ... Amor Dolus †momus Labor Invidentia Fatum Senectus Mors Tenebrae Miseria Querella Gratia Fraus Pertinacia Parcae Hesperides Somnia; quos omnis Erebo et Nocte natos ferunt.

> [... Love, Trickery, Woe, Toil, Envy, Fate, Old Age, Death, Shadows, Sadness, Quarrels, Graces, Fraud, Obstinacy, Parcae, Hesperides, Dreams, and all of whom they say were born from Darkness and Night.]

Hyginus's very similar list follows (*Preface*):

> ex Nocte et Erebo Fatum Senectus Mors Letum †Continentia Somnus Somnia <Amor> id est Lysimeles, Epiphron †dumiles Porphyrion Epaphus Discordia Miseria Petulantia Nemesis Euphrosyne Amicitia Misericordia Styx; Parcae tres, id est Clotho Lachesis Atropos; Hesperides, Aegle Hesperie †aerica.

> [From Night and Darkness, Fate, Old Age, Death, Oblivion, Continence (?), Sleep, Dreams, Love, i.e., The Limb-Loosener, Epiphron, dumiles (?) Porphyrion, Epaphus, Strife, Sadness, Petulance, Nemesis, Kindness, Friendship, Misery, Styx, the three Parcae, i.e., Clotho, Lachesis, Atropos, Hesperides, Aegle, Hesperia, Aerica.]

The most likely places where Cicero and Hyginus may have translated Hesiod's Keres occur in the respective sequences Fatum Senectus Mors Tenebrae and Fatum Senectus Mors Letum. Arthur Stanley Pease (1968) has cited a portion of Hesiod's catalogue of Night from the *Theogony* as a likely parallel (*Th.* 211–12; 217): Νὺξ δ'ἔτικτεν στυγερόν τε Μόρον καὶ Κῆρα μέλαιναν καὶ Θάνατον; καὶ Μοίρας καὶ Κῆρας ἐγείνατο νηλεοποίνους. These parallels suggest that perhaps Tenebrae or Letum may have been translations of Keres. More crucially, the term dirae does not occur in either catalogue. Therefore, insofar as the Latin lists can be viewed as translations of Hesiod, even loose ones, it seems unlikely that dirae would be chosen for Keres as Coxon had previously suggested.

Fragment 18 and the Erinyes

To return to the Erinyes possibility, it is important to recognize the fact that, in Hesiod, the Erinyes are *not* members of the Catalogue of Night. There are, however, other traditions, as in Aeschylus's *Eumenides,* that suggest their inclusion (416–17). Furthermore, there are signs in Hesiod that the poet was aware of this tradition. Most importantly, their birth in the *Theogony* occurs in an episode that interrupts the poet's list of Night's children to tell the story of strife between Gaia and Ouranos. After Cronus and Gaia work together to castrate Ouranus, the blood and genitals falling to the earth and ocean results in the birth of not only the Erinyes but also the Giants, Ash tree nymphs, and, most importantly, Aphrodite. The episode reinforces the cosmic significance of Aphrodite's birth in the *Theogony,* which constantly relies upon the interplay of love and strife (188–93):[32]

> μήδεα δ' ὡς τὸ πρῶτον ἀποτμήξας ἀδάμαντι
> κάββαλ' ἀπ' ἠπείροιο πολυκλύστῳ ἐνὶ πόντῳ,
> ὣς φέρετ' ἄμ πέλαγος πουλὺν χρόνον, ἀμφὶ δὲ λευκὸς
> ἀφρὸς ἀπ' ἀθανάτου χροὸς ὤρνυτο· τῷ δ' ἔνι κούρη
> ἐθρέφθη·

> [When first he reaped the genitals with the adamant and
>   tossed them down from the land
> into the stormy sea, as the sea carried them for a long time,
>   and around them a white
> foam arose from the immortal flesh, and in the foam a
>   maiden congealed.]

Even if this scene describes an anomalous or aberrant birth, like Parmenides's Fragment 18, it could be considered the earliest example of embryology in epic poetry. The vividness of the verb ἐθρέφθη stems from the etymology of τρέφω, which means "to thicken" (Demont 1978). The verb frequently describes the curdling of milk to make cheese, as at *Odyssey* 9.246, and so, along with πήγνυμι and συνίστημι, the verb also suggests the natural process by which the liquid semen is "set" and hardened into a baby.³³ Also, like Parmenides's fragment, Hesiod includes the constellation of Eros, strife between parents, and the Erinyes, who, just before this moment, are the first to be born from the Earth receiving the blood of Ouranos (*Theogony* 183–87):³⁴

ὅσσαι γὰρ ῥαθάμιγγες ἀπέσσυθεν αἱματόεσσαι,
πάσας δέξατο Γαῖα· περιπλομένων δ' ἐνιαυτῶν
γείνατ' Ἐρινῦς τε κρατερὰς μεγάλους τε Γίγαντας,
τεύχεσι λαμπομένους, δολίχ' ἔγχεα χερσὶν ἔχοντας,
Νύμφας θ' ἃς Μελίας καλέουσ' ἐπ' ἀπείρονα γαῖαν.

[For as many as the bloody drops that flew off,
The earth received them all, and with the years revolving
She bore both the strong Erinyes and great Giants,
Gleaming in their armor, holding long spears in their hands,
And the Nymphs, whom they call Meliai on the boundless earth.]

Furthermore, it is after this scene that Hesiod picks up his list of Night's children once again, part of this list mentioned above. As traditional children of Night, the Erinyes's birth helps justify Hesiod's own interruption of her catalogue. The Hesiodic passage also gives the Erinyes an important connotation with birth, which justifies their presence in Fragment 18 of Parmenides.

I would like to suggest another potential model for Parmenides's embryological fragment. In *Works and Days*, Hesiod concludes the poem with the story of the birth of Oath from Strife, the youngest child of Night's lineage (801–4):

Πέμπτας δ' ἐξαλέασθαι, ἐπεὶ χαλεπαί τε καὶ αἰναί·
ἐν πέμπτῃ γάρ φασιν Ἐρινύας ἀμφιπολεύειν
Ὅρκον γεινόμενον, τὸν Ἔρις τέκε πῆμ' ἐπιόρκοις.

[Avoid fifth-days, since they are difficult and dreadful.
For they say that on the fifth day the Erinyes attended
Oath as he was born, whom Eris gave birth to as a bane to
  those who swear falsely.]

Like Coxon's suggestion mentioned above, the participle "being born," γεινόμενον, matches the nascentem of Caelius. Also, the Erinyes tending to the birth in Hesiod is similar to their harassing the sex of the child. Although "to harass" and "to tend" seem opposite, there is an example from Homer in which ἀμφιπολεύειν is used euphemistically to describe the captives of the Erinyes serving them as handmaids.[35] There is also a thematic connection, since both Strife and Oath—a kind of harmony—reappear in Parmenides's fragment as the strife and harmony of the parents' seeds. Since Parmenidean embryology is frequently related to Alcmaeon's theories, it is also noteworthy that the harmony of Fragment 18—perhaps a physical analogue to the oath of Hesiod's *Theogony*—is functionally equivalent to Alcmaeon's notion of ἰσονομία as a model for health (B30), which describes the balance of opposite forces.

As is often observed, since Hesiod's *Theogony* is arranged genealogically, it makes sense that Eros would be among the first principles of his cosmos. That Parmenides chooses to do the same casts a cosmic glow over all his embryological theories. What occurs in every particular act of human procreation matches what happens on the cosmic level, as the goddess's activities and Eros suggest. In Hesiod, the Erinyes also symbolize the cosmic significance of every act of procreation. Parmenides included them not only in imitation of Hesiod but to say that reproduction, like cosmogony, follows certain principles. The Erinyes are there to ensure that natural order is maintained.[36] Their appearance in Heraclitus follows a similar pattern:[37]

κατὰ [....]ᾳ Ἡράκλειτος μα[..........] τὰ κοινά, κατ [αστρέ]φει τὰ
ἴδ[ι]α, ὅσπερ ἴκελᾳ [τῶι ἱερο]λόγωι λέγων [
"ἥλι[ος περιο]δου κατὰ φύcιν ἀγθρω[πηΐου] εὖρος ποδός [ἐcτι,
τὸ μ[έγεθο]c οὐχ ὑπερβάλλων εἰκ[ότας οὔ]ρους ε[ὔρους
ἑοῦ· εἰ δὲ μ]ή, Ἐρινύε[c] νιν ἐξευρήcου[cι, Δίκης ἐπίκουροι.

[Heraclitus . . . common things. . . . turns his own views upside down, the one who said, speaking like the Hierologos: The sun, in the nature of a circuit(?) is the breadth of a human foot, not overstepping in size the proper limits of its width, or else the Erinyes, Dike's assistants, will find him out.]

Heraclitus calls the Erinyes the assistants of Dike, a goddess who also plays an important role in Parmenides. As Dike's assistants, the Erinyes are agents maintaining cosmic balance. Also relevant is that, in Heraclitus, the Erinyes threaten to pursue the Sun if he oversteps his bounds. In Parmenides, it is the daughters of the sun who guide the kouros in his chariot to encounter Dike (Fragment 1). Since the Erinyes are daughters of Night, they are effectively the opposite of the Heliades of Parmenides's proem. Perhaps the connection between the Heliades and Dike in the Proem is enough to suggest that Parmenides would eventually also include Dike's assistants, the Erinyes.

Conclusion

After Hesiod, the Erinyes continue to be mentioned frequently in the context of familial relationships between parents and children. Consider the scene at the end of the *Libation Bearers*. After Orestes murders Clytemnestra, he is immediately tormented by the Erinyes, who here represent not only divine vengeance, but also the guilt Orestes feels for murdering his own mother. It is appropriate that in the conclusion to the trilogy, Apollo should question the blood relationship between Orestes and Clytemnestra; but too little attention is paid to the exchange between the Erinyes and Orestes before Apollo's famous defense (*Eu.*, 606–8):

> Χο. οὐκ ἦν ὅμαιμος φωτὸς ὃν κατέκτανεν.
> Ορ. ἐγὼ δὲ μητρὸς τῆς ἐμῆς ἐν αἵματι;
> Χο. πῶς γάρ σ' ἔθρεψεν ἐντός, ὦ μιαιφόνε,
> ζώνης; ἀπεύχηι μητρὸς αἷμα φίλτατον;

> [Eumenides: She [sc. Clytemnestra] was not of the same blood as the man she killed.
> Orestes: But do I share in the blood of my mother?
> Eumenides: Otherwise, how did she nourish you within her womb? Do you reject the dearest blood of your mother?]

Although it is often stated that Apollo's argument represents the "traditional" view of conception, one can argue that, in the scope of the *Oresteia*, the

Eumenides represent tradition as the elder gods when compared with the young Apollo, arguing against them, and Athena, who ultimately tries to appease them. It is, therefore, important to recognize that they seem to believe that the mother is not merely a vessel, as Apollo says. Furthermore, Athena's anomalous birth from the head of Zeus cannot be cited as a paradigm for human embryonic development. Apollo's view, therefore, is one of many, while the Erinyes accept the mother's contribution, and are more compatible with the theory of conception in Parmenides.

In conclusion, since Caelius's Latin has only problematic connections to Parmenides's philosophy, perhaps it is time to throw his interpretation of Fragment 18 out entirely. It is possible that Fragment 18 was not about the sex, gender, or even the character of the child. Instead, it was a theory about successful and unsuccessful conception that took the harmony between the physical parents' seeds as its most vital requirement. It is possible that, as agents of maintaining order and allies of Dike and Necessity, the Erinyes prevent children from being born if the unification of disparate elements was not achieved by the parents. This is not to punish the unborn, but to emphasize the necessity of Parmenides's "two-seed" theory and, in turn, the two principles of his cosmos: light and darkness.

Notes

1. Scil. Dirae (Drabkin, cf. Schrijvers 1985).
2. Translations are my own unless otherwise noted.
3. For Anaximander, see Baldry 1932; Kahn 1994, 110 n.1, 156 n.1; For Thales, see Aristotle *Met.* A3 983b22–27 = Thales A12 (DK).
4. Marciano 2005; Wilford 1968.
5. Schrijvers 1985; Coxon 2009; Mansfeld 2018; Cherubin 2019.
6. Sexus, perhaps derived from seco, "to cut," implying a stricter division between the sexes and has no Greek analogue, unlike genus, which does (γένος). In Latin, genus is the term for (grammatical) gender, while sexus denotes biological sex (Corbeill 2015, 5–8).
7. Schrijvers 1985, 45, reports that C.J. Ruijgh (Amsterdam) suggested the line ending ἐν σώματ' Ἐρίνυς [sic], ". . . in the body, an Erinys . . ."
8. See esp. Journée 2012.
9. This may be an allusion to Atlas, based on the parallel in Hesiod that Coxon suggests (1986, 69, cf. Hesiod *Th.* 517–18).
10. For a survey of biological—"vitalist"—analogies, see Lloyd 1966, 232–72. For connections to Hesiod's catalogue of Night, see Northrup 1980.

11. For Hesiod's influence on Parmenides see, e.g., Pellikaan-Engel 1978, 8–10, for a list of verbal parallels between the two authors.

12. The extent of Caelius's dependence on Soranus is a matter of debate among scholars, for which see van der Eijk 1999 and 2005, with bibliography.

13. Wilamowitz-Moellendorff 1913, 72 n. 1; Coxon 2009 [1986], 253.

14. Diels 1897, 115–18; Lesky 1950, 47–50; Untersteiner 1958; Hölscher 1986.

15. Lesky 1950, 49. See also Cherubin (2019, 36 n. 15 and 39–40), who argues this could explain how individuals obtain various configurations of both primary and secondary sex characteristics.

16. This conflates a variety of different criticisms; see Cherubin 2019, 29–36 with bibliography.

17. Shrijvers (1985, 22–25) suspects the prejudice is due to Christian or Stoic influence on Caelius (cf. van der Eijk 2005, 302).

18. Many scholars believe Parmenides adopted embryological views of the Pythagoreans, like Alcmaeon; but more recent arguments suggest Alcmaeon was not a predecessor, but later than Parmenides (Mansfeld 2018 [2015], n. 36).

19. Alcmaeon A13, A14 (DK).

20. Caelius and Soranus are representatives of the "Methodist" school of medicine, which sought to teach one practical "method" of medicine for the treatment of all diseases (Touwaide 2006).

21. It is notable that the phrase does not occur in Lucretius's discussions of sexual intercourse (e.g., 4.1030–57, 1209–73).

22. Text Heyworth 2007. This parallel was pointed out by C. J. Ruijgh to Schrijvers, presumably in a personal conversation, as no other source is cited by Schrijvers (1985, 45).

23. There are problems with this translation, since the intransitive is not found in the present active forms of this verb (s.v. μείγνυμι LSJ passim). Alternatively, Diels may have intended to make δύναμις the subject of μίσγη, translated as "whenever a power mixes male and female seeds." The delayed subject is not ideal, but not impossible.

24. Alternatively, we can interpret the phrase praeter materias to mean not "beyond" but "in addition to matter," translating: "for, in addition to the matter of the seeds, he wants them to have powers." My thanks to an anonymous reviewer for making this suggestion. The reviewer adds that, thanks to the alternative translation, the dualist implications vanish; but I think that Caelius has already established his own dualist stance in his initial description of homosexuality as "not because of Nature" (non . . . ex natura) and "foreign to diseases of the body" (a passionibus corporis aliena, quoted above). Caelius then attempts to make the Parmenides passage fit his initial dualist interpretation of homosexuality as a psychological, and not a physical, disease.

25. Cf. Hom.*Il.*15.204, Iris's warning Poseidon not to offend Zeus: "You know how the Erinyes always attend to the older [sc. siblings]," (οἶσθ' ὡς πρεσβυτέροισιν Ἐρινύες αἰὲν ἕπονται).

26. The corresponding pairs are Night/Light; Masculine/Feminine; Cold/Hot; Dense/Rare; Right/Left, etc. There is some disagreement about whether masculine or feminine should be the cold, dark one. The pairs as I list them agree with Journée 2012, pace Guthrie 1965, 77–80.

27. On this issue, see Michael Wiitala's chapter in this volume.

28. This statistic is based on the author's TLG search.

29. Plutarch adds that they mix to form all things (Plut. *adv. Col.* 1114 B=B10 DK).

30. Why Parmenides describes birth this way is unknown. A recent suggestion is that it could be due to the dangers birth poses to both mother and child, as there is no guarantee of survival for either of them, especially in antiquity (Cherubin 2019, 34). Whatever is meant, the negative valence matches what is seen in Fragment 18, the fighting of the virtues and the resulting "appetites" that pursue the child. It is even possible that the term dirae from Fragment 18, if originally intended as an adjective, translates στυγεραί.

31. See Coxon 2009 on this fragment's debate. Coxon argues the subject is Genesis herself.

32. Bonnafé 1985, 28–30. It may also be significant for Parmenides that the Scholion to *Th.* 187 claims humans came from Giants and Melian Nymphs united (cf. Strauss Clay 2003, 95–99).

33. See, for example, Sur les femmes steriles Littré v.8, 412.

34. West (1966) makes the connection between this passage and later hematogenous theories of conception, such as Diogenes of Apollonia B6 (DK).

35. Hom.*Od.*20.78: καί ῥ' ἔδοσαν στυγερῇσιν Ἐρινύσιν ἀμφιπολεύειν· "And they gave them to the Erinyes to serve . . ." The Greek verb ἕπομαι, like sequor in Caelius's paraphrase (4.9.135, quoted above), bears a similar ambiguity and appropriately describes the Erinyes's activities.

36. For the Erinyes as protectors of natural order, see Roscher 1884, 1323 (cf. Schrijvers 1985, 68 n. 85). For their appearance in Heraclitus in this role, see Marcovich 2001, 274–77.

37. This fragment is similar to fr. 94 (DK), whose original source was Plutarch *de exil.* 604a, but it includes the added material from the Derveni Papyrus (Col. IV.5–9). Text by Tsantsanoglou, translation by Kouremenos, adapted.

Works Cited

Adluri, Vishwa. 2011. *Parmenides, Plato, and Mortal Philosophy: Return from Transcendence.* New York: Continuum.
Ahrensdorf, Peter J. 2014. *Homer on the Gods and Human Virtue: Creating the Foundations of Classical Civilization.* Cambridge: Cambridge University Press.
Alexander, Caroline, trans. 2015. *The Iliad.* New York: HarperCollins.
Altman, William H. F. 2015. "Parmenides Fragment B3 Revisited," *Hypnos* 35: 197-230.
Austin, Scott. 2007. *Parmenides and the History of Dialectic: Three Essays.* Las Vegas: Parmenides.
Baldry, H. C. 1932. "Embryological Analogies in Presocratic Cosmogony," *Classical Quarterly* 26: 27-34.
Barnes, Jonathan. 1979. "Parmenides and the Eleatic One," *Archiv für Geschichte der Philosophie* 61: 1-21.
Barnes, Jonathan. 1982. *The Presocratic Philosophers.* Revised edition. New York: Routledge.
Barrett, James. 2004. "Struggling with Parmenides," *Ancient Philosophy* 24, no. 2: 267-91.
Benardete, Seth. 1997. *The Bow and the Lyre: A Platonic Reading of the* Odyssey. Lanham, MD: Rowman & Littlefield.
Benzi, Nicolò. 2016. "Noos and Mortal Enquiry in the Poetry of Xenophanes and Parmenides," *Methodos. Savoirs et Textes* 16: 1-18.
Bolotin, David. 1989. "The Concerns of Odysseus: An Introduction to the *Odyssey*," *Interpretation: A Journal of Political Philosophy* 17: 41-57.
Bonnafé, Annie. 1985. *Eros et Eris: Mariages divins et mythe de succession chez Hésiode.* Lyon: Presses universitaires de Lyon.
Bowra, C. M. 1937. "The Proem of Parmenides," *Classical Philology* 32, no. 2: 97-112.
Brann, Eva. 2001. *The Ways of Naysaying: No, Not, Nothing, and Nonbeing.* Lanham, MD: Rowman & Littlefield.
Brann, Eva, Peter Kalkavage, and Eric Salem, trans. 1996. *Plato's Sophist or the Professor of Wisdom.* Newburyport, MA: Focus.

Bredlow, Luis Andrés. 2011. "Parmenides and the Grammar of Being," *Classical Philology* 106, no. 4: 283–98.
Brown, Lesley. 1986. "Being in the *Sophist*: A Syntactical Enquiry," *Oxford Studies in Ancient Philosophy* 4: 49–70.
Brown, Lesley. 1994. "The Verb 'to Be' in Greek Philosophy." In *Language*, edited by Stephen Everson, 212–37. Cambridge: Cambridge University Press.
Bryan, Jenny. 2012. *Likeness and Likelihood in the Presocratics and Plato*. Cambridge: Cambridge University Press.
Bryan, Jenny. 2018. "Reconsidering the Authority of Parmenides' Doxa." In *Authors and Authorities in Ancient Philosophy*, edited by Jenny Bryan, Robert Wardy, and James Warren, 1–19. Cambridge: Cambridge University Press.
Bryan, Jenny. 2020. "The Pursuit of Parmenidean Clarity," *Rhizomata* 8: 218–38.
Burkert, Walter. 1969. "Das Proömium des Parmenides und die 'Katabasis' des Pythagoras," *Phronesis* 14, no. 1: 1–30.
Burkert, Walter. 1987. *Ancient Mystery Cults*. Cambridge, MA: Harvard University Press.
Burkert, Walter. 2012. *Greek Religion: Archaic and Classical*. Malden, MA: Blackwell.
Burnet, John. 1930. *Early Greek Philosophy*. 4th ed. London: Adam and Charles Black.
Calogero, Guido. 1932. *Studi sull' Eleatismo*. Rome: La Nuova Italia.
Carson, Anne, trans. 2003. *If Not, Winter: Fragments of Sappho*. New York: Random House.
Casertano, Giovanni. 2011. "Parmenides—Scholar of Nature." In *Parmenides, Venerable and Awesome*, edited by Néstor-Luis Cordero, 21–58. Las Vegas: Parmenides.
Cerri, Giovanni. 2011. "The Astronomical Section in Parmenides' Poem." In *Parmenides, Venerable and Awesome*, edited by Néstor-Luis Cordero, 81–94. Las Vegas: Parmenides.
Cherubin, Rose. 2004. "Parmenides' Poetic Frame," *International Studies in Philosophy* 36, no. 1: 7–38.
Cherubin, Rose. 2005. "Light, Night, and the Opinions of Mortals: Parmenides B8.51–61 and B9," *Ancient Philosophy* 25, no. 1: 1–23.
Cherubin, Rose. 2017. "Mortals Lay Down Trusting to be True," *Epoché* 21, no. 2: 251–71.
Cherubin, Rose. 2019. "Sex, Gender, and Class in the Poem of Parmenides: Difference without Dualism?" *American Journal of Philology* 140, no. 1: 29–66.
Cole, Susan Guettel. 1984. "The Social Function of Rituals of Maturation: The Koureion and the Arkteia," *Zeitschrift für Papyrologie und Epigraphik* 55: 233–44.
Corbeill, Anthony. 2015. *Sexing the World: Grammatical Gender and Biological Sex in Ancient Rome*. Princeton, NJ: Princeton University Press.
Cordero, Néstor-Luis. 1979. "Les deux chemins de Parménide dans les fragments 6 et 7," *Phronesis* 24: 1–32.

Cordero, Néstor-Luis. 1984. *Les deux chemins de Parménide*. Paris-Bruxelles: Vrin-Ousia.
Cordero, Néstor-Luis. 2004. *By Being, It Is: The Thesis of Parmenides*. 2nd ed. Las Vegas: Parmenides.
Cordero, Néstor-Luis. 2010. "The Doxa of Parmenides Dismantled," *Ancient Philosophy* 30, no. 2: 231–46.
Cordero, Néstor-Luis, ed. 2011. *Parmenides, Venerable and Awesome: Proceedings of the International Symposium*. Las Vegas: Parmenides.
Cordero, Néstor-Luis. 2017. "La place de la « physique » de Parménide dans une nouvelle reconstitution du Poème," *Revue de philosophie ancienne* 35, no. 1: 3–13.
Cornford, Francis M. 1933. "Parmenides' Two Ways," *Classical Quarterly* 27, no. 2: 97–111.
Cornford, Francis M. 1935. "A New Fragment of Parmenides," *Classical Review* 49, no. 4: 122–23.
Cornford, Francis M. 1939. *Plato and Parmenides*. London: Routledge and Kegan Paul.
Cosgrove, Matthew R. 1974. "The 'KOYROS' Motif in Parmenides: B 1.24," *Phronesis* 19: 81–94.
Cosgrove, Matthew R. 2011. "The Unknown 'Knowing Man': Parmenides, B1.3," *Classical Quarterly* 61, no. 1: 28–47.
Cosgrove, Matthew R. 2014. "What Are 'True' Doxai Worth to Parmenides? Essaying a Fresh Look at His Cosmology," *Oxford Studies in Ancient Philosophy* 46: 1–32.
Coxon, Allan H. 2003. "Parmenides on Thinking and Being," *Mnemosyne* 56, no. 2: 210–12.
Coxon, Allan H. 2009. *The Fragments of Parmenides*. Revised and expanded edition by Richard McKirahan. Las Vegas: Parmenides.
Curd, Patricia. 2004. *The Legacy of Parmenides: Eleatic Monism and Later Presocratic Thought*. 2nd ed. Las Vegas: Parmenides.
Curd, Patricia, ed.; Richard McKirahan and Patricia Curd, trans. 2011. *A Presocratics Reader: Selected Fragments and Testimonia*. 2nd ed. Indianapolis: Hackett.
Curd, Patricia. 2015. "Thinking, Supposing, and Physis in Parmenides," *Études Platoniciennes* 12 (online).
De Rijk, L. M. 1983. "Did Parmenides Reject the Sensible World?" In *Graceful Reason*, edited by Lloyd P. Gerson, 29–53. Toronto: Pontifical Institute of Mediaeval Studies.
De Vaan, Michiel. 2008. *Etymological Dictionary of Latin and the Other Italic Languages*. Leiden: Brill.
DeLong, Jeremy C. 2015. "Rearranging Parmenides: B1: 31–32 and a Case For an Entirely Negative Doxa (Opinion)," *Southwest Philosophy Review* 31, no. 1: 177–86.
DeLong, Jeremy C. 2016. "Parmenides' Theistic Metaphysics." PhD diss., University of Kansas.

DeLong, Jeremy C. 2017. "From Ionian Speculation to Eleatic Deduction: Parmenides' Xenophanean-Based Theism," In *Politics and Performance in Western Greece*, edited by Heather Reid, 221–36. Sioux City: Parnassos Press.

DeLong, Jeremy C. 2018a. "Parmenides of Elea." In *Internet Encyclopedia of Philosophy*, edited by James Fieser and Bradley Dowden.

DeLong, Jeremy C. 2018b. "Parmenides, Plato, and Μίμησις." In *The Many Faces of Mimesis*, edited by Heather L. Reid and Jeremy C. DeLong, 61–74. Sioux City: Parnassos Press.

Denniston, J. D. 1996. *The Greek Particles*. 2nd ed. Indianapolis: Hackett.

Detienne, Marcel. 1996. *Masters of Truth in Archaic Greece*. Translated by Janet Lloyd. New York: Urzone, 1996.

Detienne, Marcel, and Jean Pierre Vernant. 1991. *Cunning Intelligence in Greek Culture and Society.* Translated by Janet Lloyd. Chicago: University of Chicago Press.

Diels, Hermann. 1897. *Parmenides Lehrgedicht: Griechisch und Deutsch*. Berlin: Reimer.

Diels, Hermann and Walther Kranz (ed.). 1956. *Die Fragmente der Vorsokratiker*. Vol. 1. Zürich-Berlin: Weidmannsche Verlagsbuchhandlung.

Diller, Hans. 1946. "Hesiod und die Anfänge der griechischen Philosophie," *Antike und Abendland II*: 140–51.

Dodds, E. R. 1951. *The Greeks and the Irrational.* Berkeley: University of California Press.

Drabkin, Israel E., trans. 1951. *Caelius Aurelianus: On Acute and On Chronic Diseases.* Chicago: University of Chicago Press.

Ebert, Theodore. 1983. "Aristotle on What Is Done In Perceiving," *Zeitschrift für philosophische Forschung* 37, no. 2: 181–98.

Elbert Decker, Jessica. 2019. "How to Speak Kata Physin: Magicoreligious Speech in Heraclitus," *Epoché* 23, no. 2: 263–74.

Elbert Decker, Jessica. 2021. "I Will Tell a Double Tale: Double Speak in the Ancient Greek Poetic Tradition," *Epoché* 25, no. 2: 237–48.

Eliade, Mircea. 1958. *Birth and Rebirth: The Religious Meanings of Initiation in Human Culture.* 1st ed. New York: Harper and Row.

Empson, William. 1966. *Seven Types of Ambiguity.* Revised from British 1930 edition. New York: New Directions.

Evans, Nancy A. 2002. "Sanctuaries, Sacrifices, and the Eleusinian Mysteries," *Numen* 49, no. 3: 227–54.

Evelyn-White, Hugh G. 1914. *The Homeric Hymns and Homerica with an English Translation.* Cambridge, MA: Harvard University Press.

Finkelberg, Aryeh. 1986. "The Cosmology of Parmenides," *American Journal of Philology* 107, no. 3: 303–17.

Frede, Dorothea. 1993. "The Question of Being: Heidegger's Project." In *The Cambridge Companion to Heidegger*, edited by Charles Guignon, 42–69. Cambridge: Cambridge University Press.

Friedman, Jane. 2020. "The Epistemic and the Zetetic," *Philosophical Review* 129, no. 4: 501–36.

Fronterotta, Francesco. 2007. "Some Remarks on Noein in Parmenides." In *Reading Ancient Texts. Volume I: Presocratics and Plato: Essays in Honour of Denis O'Brien*, edited by Suzanne Stern-Gillet and Kevin Corrigan, 12–18. Boston: Brill.

Fülleborn, Georg Gustav. 2012. *Fragmente des Parmenides*. Reprint of 1795 original. Charleston, SC: Nabu Press.

Furth, Montgomery. 1968. "Elements of Eleatic Ontology," *Journal of the History of Philosophy* 6, no. 2: 11–132.

Gallop, David. 1984. *Parmenides of Elea: Fragments: A Text and Translation with an Introduction*. Toronto: University of Toronto Press.

Giancola, Donna M. 2001. "Toward a Radical Reinterpretation of Parmenides' B3," *Journal of Philosophical Research* 26: 635–53.

Gonzalez, Francisco. 2019. "Being as Activity: A Defence of the Importance of *Metaphysics* 1048B18–35 for Aristotle's Ontology," *Oxford Studies in Ancient Philosophy* 56: 123–92.

Graham, Daniel W. 2002. "La Lumière de la lune dans la pensée grecque archaïque." In *Qu'est-ce que la Philosophie Présocratique*, edited by André Laks and Claire Louguet. Villeneuve d'Ascq: Presses Universitaires du Septentrion.

Graham, Daniel W. 2006. *Explaining the Cosmos: The Ionian Tradition of Scientific Philosophy*. Princeton, NJ: Princeton University Press.

Graham, Daniel W., ed. 2010. *The Texts of Early Greek Philosophy: The Complete Fragments and Selected Testimonies of the Major Presocratics: Part I*. Cambridge: Cambridge University Press.

Graham, Daniel W. 2013. *Science Before Socrates: Parmenides, Anaxagoras, and the New Astronomy*. Oxford: Oxford University Press.

Granger, Herbert. 2010. "Parmenides of Elea: Rationalist or Dogmatist," *Ancient Philosophy* 30, no. 1: 15–38.

Gregory, Andrew. 2014. "Parmenides, Cosmology and Sufficient Reason," *Apeiron* 47, no. 1: 16–47.

Guthrie, W. K. C. 1962. *A History of Greek Philosophy, Volume 1: The Earlier Presocratics and the Pythagoreans*. Cambridge: Cambridge University Press.

Guthrie, W. K. C. 1965. *A History of Greek Philosophy, Volume 2: The Presocratic Tradition from Parmenides to Democritus*. Cambridge: Cambridge University Press.

Haas, Jens, and Katja Maria Vogt. 2020. "Incomplete Ignorance." In *Epistemology After Sextus Empiricus*, edited by Katja Maria Vogt and Justin Vlasits, 254–67. Oxford: Oxford University Press.

Havelock, Eric A. 1958. "Parmenides and Odysseus," *Harvard Studies in Classical Philology* 63: 133–43.

Heath, Thomas. 1912. *Aristarchus of Samos*. Oxford: Oxford University Press.
Heidegger, Martin. 1935. *Introduction to Metaphysics*. New Haven, CT: Yale University Press.
Heidegger, Martin. 1975. *Early Greek Thinking*. Translated by David Farrell Krell and F. A. Capuzzi. San Francisco: Harper.
Heidegger, Martin. 1976. *Hölderlins Hymne »Der Ister«*. Vol. 53. Gesamtausgabe. Frankfurt am Main: Klostermann.
Henn, Martin J. 2003. *Parmenides of Elea: A Verse Translation with Interpretative Essays and Commentary to the Text*. Westport, CT: Praeger.
Hintikka, Jakko. 1980. "Parmenides' Cogito Argument," *Ancient Philosophy* 1: 5–16.
Hölscher, Uvo. 1986. *Parmenides: Vom Wesen des Seienden*. Berlin: Suhrkamp.
Horster, Marietta. 2010. "Religious Landscape and Sacred Ground: Relationships between Space and Cult in the Greek World," *Revue de l'histoire des religions* 227, no. 4: 435–58.
Inwood, Brad. 2001. *The Poem of Empedocles: A Text and Translation with an Introduction*. Rev. ed. Toronto: University of Toronto Press.
Johansen, Thomas. 2016. "Parmenides' Likely Story," *Oxford Studies in Ancient Philosophy* 50: 1–29.
Johnston, Sarah Iles. 2006. "Erinyes." In *Brill's New Pauly*. Leiden: Brill.
Journée, Gérard. 2012. "Lumière et Nuit, Féminin et Masculin chez Parménide d'Elée: quelques remarques," *Phronesis* 57, no. 4: 289–318.
Kahn, Charles. 1966. "The Greek Verb 'to Be' and the Concept of Being," *Foundations of Language* 2: 245–65.
Kahn, Charles. 1969. "The Thesis of Parmenides," *Review of Metaphysics* 22, no 4: 700–24.
Kahn, Charles. 1979. *The Art and Thought of Heraclitus: An Edition of the Fragments with Translation and Commentary*. Cambridge: Cambridge University Press.
Kahn, Charles. 1994. *Anaximander and the Origins of Greek Cosmology*. Indianapolis: Hackett.
Kahn, Charles. 2003. *The Verb 'Be' in Ancient Greek*. Indianapolis: Hackett.
Kahn, Charles. 2012. *Essays on Being*. Oxford: Oxford University Press.
Kelp, Christoph. 2014. "Two for the Knowledge Goal of Inquiry," *American Philosophical Quarterly* 51, no. 3: 227–32.
Kember, Owen. 1971. "Right and Left in the Sexual Theories of Parmenides," *The Journal of Hellenic Studies* 91: 70–79.
Ketchum, Jonathan. 1981. *The Structure of the Plato Dialogue*. PhD diss., State University of New York at Buffalo.
Ketchum, Richard. 1990. "Parmenides on What There Is," *Canadian Journal of Philosophy* 20, no. 2: 167–90.
Kingsley, Peter. 1999. *In the Dark Places of Wisdom*. Inverness, CA: Golden Sufi Press.
Kingsley, Peter. 2020. *Reality*. 2nd ed. London: Catafalque Press.

Kirk, G. S., J. E. Raven, and M. Schofield. 1983. *The Presocratic Philosophers*. 2nd ed. Cambridge: Cambridge University Press.

Kosman, Aryeh. 2013. *The Activity of Being*. Cambridge, MA: Harvard University Press.

Kouremenos, Theokritos, George M. Parássoglou, and Kyriakos Tsantsanoglou. 2006. *The Derveni Papyrus*. Florence: Casa Editrice.

Kurfess, Christopher. 2012. "Restoring Parmenides' Poem: Essays toward a New Arrangement of the Fragments Based on a Reassessment of the Original Sources." PhD diss., University of Pittsburgh.

Kurfess, Christopher. 2016. "The Truth about Parmenides' Doxa," *Ancient Philosophy* 36, no. 1: 13–45.

Laks, André, and Glenn W. Most. 2016. *Early Greek Philosophy: Western Greek Thinkers. Part 2*. Cambridge, MA: Harvard University Press.

Lennox, James G. 2021. *Aristotle on Inquiry: Erotetic Frameworks and Domain-Specific Norms*. Cambridge: Cambridge University Press.

Lesher, James H. 1981. "Perceiving and Knowing in the *Iliad* and *Odyssey*," *Phronesis* 26, no. 1: 2–24.

Lesher, James H. 1984. "Parmenides' Critique of Thinking," *Oxford Studies in Ancient Philosophy* 2: 1–31.

Lesher, James H. 1992. *Xenophanes of Colophon: A Text and Translation with a Commentary*. Toronto: University of Toronto Press.

Lesher, James H. 2002. "Parmenidean Elenchos." In *Does Socrates Have a Method? Rethinking the Elenchus in Plato's Dialogues and Beyond*, edited by Gary Alan Scott, 19–35. University Park: Pennsylvania State University Press.

Lesher, James H. 2008. "The Humanizing of Knowledge in Presocratic Thought." In *The Oxford Handbook of Presocratic Philosophy*, edited by Patricia Curd and Daniel W. Graham, 458–84. New York: Oxford University Press.

Lesky, Erna. 1950. *Die Zeugungs- und Vererbungslehre der Antike und ihr Nachwirken*. Mainz: Verlag der Akademie der Wissenschaften und der Literatur.

Lewis, Frank. 2009. "Parmenides' Modal Fallacy," *Phronesis* 54, no. 1: 1–8.

Lloyd, G. E. R. 1966. *Polarity and Analogy: Two Types of Argumentation in Early Greek Thought*. Cambridge: Cambridge University Press.

Lloyd-Jones, Hugh. 1994. *Sophocles. Oedipus Tyrannus*. Loeb Classical Edition. Cambridge, MA: Harvard University Press.

Long, A. A. 1975. "The Principles of Parmenides' Cosmogony." In *Studies in Presocratic Philosophy 2: The Eleatics and the Pluralists*, edited by R. E. Allen and David J. Furley, 82–101. New York: Routledge

Long, A. A. 1996. "Parmenides on Thinking Being," *Proceedings of the Boston Area Colloquium in Ancient Philosophy* 12: 125–51.

Manchester, Peter. 2005. *The Syntax of Time: The Phenomenology of Time in Greek Physics and Speculative Logic from Iamblichus to Anaximander*. Boston: Brill.

Mansfeld, Jaap. 1964. *Die Offenbarung des Parmenides und die menschliche Welt*. Assen: Van Gorcum.
Mansfeld, Jaap. 1995. "Insight by Hindsight: Intentional Unclarity in Presocratic Proems," *Bulletin of the Institute of Classical Studies* 40: 228–32.
Mansfeld, Jaap. 2018. "Parmenides from Right to Left." In *Studies of Ancient Greek Philosophy*. Leiden: Brill.
Mansfeld, Jaap, and Oliver Primavesi. 2011. *Die Vorsokratiker*. Stuttgart: Reclam.
Marciano, M. Laura Gemelli. 2005. "Empedocles' Zoogony and Embryology." In *The Empedoclean Kosmos: Structure, Process and the Question of Cyclicity. Proceedings of the Symposium Philosophiae Antiquae Tertium Myconense July 6th–July 13th, 2003. Part 1: Papers*, edited by Apostolos L. Pierris, 373–404. Patras: Institute for Philosophical Research.
Marciano, M. Laura Gemelli. 2008. "Images and Experience: At the Roots of Parmenides' Aletheia," *Ancient Philosophy* 28, no. 1: 21–48.
Marcovich, Miroslav. 2001. *Heraclitus*. Sankt Augustin: Academia Verlag.
McKirahan, Richard D. 2008. "Signs and Arguments in Parmenides B8." In *The Oxford Handbook of Presocratic Philosophy*, edited by Patricia Curd and Daniel W. Graham, 189–229. Oxford: Oxford University Press.
McKirahan, Richard D. 2010. *Philosophy Before Socrates: An Introduction with Texts and Commentary*. 2nd ed. Indianapolis: Hackett.
Menn, Stephen. 2021. "Aristotle on the Many Senses of Being," *Oxford Studies in Ancient Philosophy* 59: 187–264.
Mill, John Stuart. 1843. *A System of Logic, Ratiocinative and Inductive*. London: Longmans.
Miller, Mitchell. 1979. "Parmenides and the Disclosure of Being," *Phronesis* 13, no. 1: 12–35.
Miller, Mitchell. 1999. "Platonic Mimesis." In *Contextualizing Classics: Ideology, Performance, Dialogue*, edited by Thomas Falkner, Nancy Felson, and David Konstan, 253–66. Lanham, MD: Littlefield.
Miller, Mitchell. 2006. "Ambiguity and Transport: Reflections on the Proem to Parmenides' Poem," *Oxford Studies in Ancient Philosophy* 30: 1–47.
Miller, Patrick Lee. 2011. *Becoming God: Pure Reason in Early Greek Philosophy*. New York: Bloomsbury.
Monro, D. B. 1882. *A Grammar of the Homeric Dialect*. Oxford: Clarendon Press.
Monro, D. B., and T. W. Allen, eds. 1920. *Homeri Opera in five volumes*. Oxford: Oxford University Press.
Morrison, J. S. 1955. "Parmenides and Er," *Journal of Hellenic Studies* 75: 59–68.
Mourelatos, Alexander P. D. 1973. "Heraclitus, Parmenides, and the Naïve Metaphysics of Things." In *Exegesis and Argument: Studies in Greek Philosophy Presented to Gregory Vlastos*, edited by E. N. Lee and Alexander P. D. Mourelatos, 16–48. New York: Humanities Press.
Mourelatos, Alexander P. D. 1999. "Parmenides and the Pluralists," *Apeiron* 32: 117–30.

Mourelatos, Alexander P. D. 2008. *The Route of Parmenides*. Rev. ed. Las Vegas: Parmenides.
Mourelatos, Alexander P. D. 2011. "Parmenides, Early Greek Astronomy, and Modern Scientific Realism." In *Parmenides, Venerable and Awesome: Proceedings of the International Symposium*, edited by Néstor-Luis Cordero, 167–90. Las Vegas: Parmenides.
Mourelatos, Alexander P. D. 2012. "'The Light of Day by Night' nukti phaos, Said of the Moon in Parmenides B14." In *Presocratics and Plato: A Festschrift at Delphi in Honor of Charles Kahn*, edited by Richard Patterson, Vassilis Karasmanis, and Arnold Hermann, 25–58. Las Vegas: Parmenides.
Mourelatos, Alexander P. D. 2018. "Review of *Early Greek Philosophy, 9 Volumes*. Loeb Classical Library, by André Laks and Glenn W. Most, 524–32." *Bryn Mawr Classical Review*.
Mourelatos, Alexander P. D. 2020. "Elements of Natural Science in the Second Part of Parmenides' Poem: Comment on Livio Rossetti's Lezione I at Eleatica 2017." In *Verso la filosofia: Nuove prospettive su Parmenide, Zenone e Melisso*, edited by N. S. Galgano, S. Giombini, and F. Marcacci, 244–50. Baden: Academia-Verlag.
Murray, A. T. 1919. *The Odyssey with an English Translation*. 2 vols. Cambridge, MA: Harvard University Press.
Murray, A. T. 1954. *The Iliad with an English Translation*. 2 vols. Cambridge, MA: Loeb Classical Library.
Mylonas, George E. 1961. *Eleusis and the Eleusinian Mysteries*. Princeton, NJ: Princeton University Press.
Nagy, Gregory. 1983. "SêMA and NÓĒSIS: Some Illustrations," *Arethusa* 16, no. ½: 35–55.
Nehamas, Alexander. 1981. "On Parmenides' Three Ways of Inquiry," *Deucalion* 33, no. 4: 97–111.
Nehamas, Alexander. 2002. "Parmenidean Being/Heraclitean Fire." In *Presocratic Philosophy: Essays in Honour of Alexander Mourelatos*, edited by Victor Caston and Daniel W. Graham, 45–64. Burlington, VT: Ashgate.
Nightingale, Andrea Wilson. 2004. *Spectacles of Truth in Classical Greek Philosophy: Theoria in Its Cultural Context*. Cambridge: Cambridge University Press.
Northrup, Mark D. 1980. "Hesiodic Personifications in Parmenides A37," *Transactions of the American Philological Association* 110: 223–32.
Onians, R. B. 1951. *The Origins of European Thought: About the Body, the Mind, the Soul, the World, Time, and Fate*. Cambridge: Cambridge University Press.
Owen, G. E. L. 1960. "Eleatic Questions," *Classical Quarterly* 10, no. 1: 84–102.
Owen, G. E. L. 1971. "Plato on Not-Being." In *Plato I*, edited by Gregory Vlastos, 223–67. Garden City, NY: Doubleday.
Owens, Joseph. 1979. "Knowledge and 'Katabasis' in Parmenides," *The Monist* 62, no. 1: 15–29.

Palmer, John. 2009. *Parmenides and Presocratic Philosophy.* Oxford: Oxford University Press.
Palmer, John. 2012. "Parmenides." In *Stanford Encyclopedia of Philosophy*, edited by Edward N. Zalta. Stanford, CA: Stanford University Press. https://plato.stanford.edu/entries/parmenides.
Parker, Robert. 1996. *Miasma: Pollution and Purification in Early Greek Religion.* Oxford: Clarendon Press.
Patzia, Michael. 2016. "Xenophanes." In *Internet Encyclopedia of Philosophy*, edited by James Fieser and Bradley Dowden. https://iep.utm.edu/xenoph/.
Pease, Arthur Stanley. 1968. *M. Tulli Ciceronis: De Natura Deorum Libri III. Libri secundus et tertius.* Darmstadt: Wiss. Buchges.
Pelletier, Francis J. 1990. *Parmenides, Plato, and the Semantics of Not-Being.* Chicago: University of Chicago Press.
Pellikaan-Engel, Maja E. 1974. *Hesiod and Parmenides: A New View on their Cosmologies and on Parmenides' Proem.* Amsterdam: Adolf M. Hakkert.
Perl, Eric. 2008. *Theophany: The Neoplatonic Philosophy of Dionysius the Areopagite.* Albany: State University of New York Press.
Perl, Eric. 2014. *Thinking Being: Introduction to Metaphysics in the Classical Tradition.* Leiden: Brill.
Popper, K. R. 1992. "How the Moon Might Throw Some of Her Light upon the Two Ways of Parmenides," *Classical Quarterly* 42, no. 1: 12–19.
Priou, Alex. 2018. "Parmenides on Reason and Revelation," *Epoché* 22, no. 2: 177–202.
Pugliese Carratelli, Giovanni. 1965. "Parmenides physikós," *La parola del Passato* 20.
Pulpito, Massimo, and Alexander P. D. Mourelatos. 2018. "Parmenides and the Principle of Sufficient Reason." In Ὁδοὶ Νοῆσαι—*Ways to Think. Essays in Honour of Néstor-Luis Cordero*, edited by Massimo Pulpito and Pilar Spangenberg, 121–41. Bologna: Diogene.
Robbiano, Chiara. 2006. *Becoming Being: On Parmenides' Transformative Philosophy.* Sankt Augustin: Academia Verlag.
Robbiano, Chiara. 2011. "What Is Parmenides' Being?" In *Parmenides, Venerable and Awesome*, edited by Néstor-Luis Cordero, 213–31. Las Vegas: Parmenides.
Robbiano, Chiara. 2016. "Being Is Not an Object," *Ancient Philosophy* 36, no. 2: 263–301.
Robbiano, Chiara. 2017. "Self or Being without Boundaries: On Śaṅkara and Parmenides." In *Universe and Inner Self in Early Indian and Early Greek Thought*, edited by Richard Seaford, 134–48. Edinburgh: Edinburgh University Press.
Roscher, Wilhelm Heinrich. 1884–1937. *Ausführliches Lexikon der griechischen und römischen Mythologie.* Leipzig: B.G. Teubner.
Rossetti, Livio. 2020. *Verso la filosofia: Nuove prospettive su Parmenide, Zenone e Melisso,* edited by N. S. Galgano, S. Giombini, and F. Marcacci, 49–168. Baden: Academia Verlag.

Russell, Bertrand. 1945. *A History of Western Philosophy*. New York: Simon and Schuster.
Sattler, Barbara. 2011. "Parmenides' System: The Logical Origins of Monism," *Proceedings of the Boston Area Colloquium in Ancient Philosophy* 26: 25–90.
Schrijvers, Piet H. 1985. *Eine medizinische Erklärung der männlichen Homosexualität aus der Antike*. Amsterdam: B. R. Grüner.
Sedley, David. 1999. "Parmenides and Melissus." In *The Cambridge Companion to Early Greek Philosophy*, edited by A. A. Long, 113–33. Cambridge: Cambridge University Press.
Seidel, George Joseph. 1964. *Martin Heidegger and the Pre-Socratics: An Introduction to His Thought*. Lincoln: University of Nebraska Press.
Smith, Colin C. 2020. "Toward a Two-Route Interpretation of Parmenidean Inquiry," *Epoché* 24, no. 2: 279–97.
Stanford, William Bedell. 1939. *Ambiguity in Greek Literature: Studies in Theory and Practice*. Oxford: Basil Blackwell.
Stokes, Michael. 1971. *One and Many in Presocratic Philosophy*. Washington, D.C.: Center for Hellenic Studies.
Strauss Clay, Jenny. 2003. *Hesiod's Cosmos*. Cambridge: Cambridge University Press.
Struck, Peter. 2004. *The Birth of the Symbol*. Princeton, NJ: Princeton University Press.
Sullivan, Shirley Darcus. 1980. "How a Person relates to νόος in Homer, Hesiod, and the Greek Lyric Poets," *Glotta* 58, bd 1/2h: 33–44.
Sullivan, Shirley Darcus. 1988. "Noos and vision: Five passages in the Greek lyric poets," *Symbolae Osloenses* 63, no. 1: 7–17.
Sullivan, Shirley Darcus. 1990. "The Psychic Term Νόος in the Poetry of Hesiod," *Glotta* 68, bd 1/2h: 68–85.
Sullivan, Shirley Darcus. 1996. "Disturbances of the Mind and Heart in Early Greek Poetry," *L'Antiquité Classique*, T. 65: 31–51.
Szemerényi, Oswald. 1969. "Etyma Latina II (7–18)." In *Studi linguistici in onore di Vittore Pisani*, 963–94. Vol. 2. Brescia: Paideia.
Tarán, Leonardo. 1965. *Parmenides: A Text with Translation, Commentary, and Critical Essays*. Princeton, NJ: Princeton University Press.
Thanassas, Panagiotis. 2007. *Parmenides, Cosmos, and Being: A Philosophical Interpretation*. Milwaukee: Marquette University Press.
Tor, Shaul. 2015. "Parmenides' Epistemology and the Two Parts of his Poem," *Phronesis* 60, no. 1: 3–39.
Tor, Shaul. 2017. *Mortal and Divine in Early Greek Epistemology*. Cambridge: Cambridge University Press.
Touwaide, Alain. 2006. "Methodists." In *Brill's New Pauly*. Leiden: Brill.
Untersteiner, Mario. 1958. *Parmenide, Testimonianze e Frammenti*. Florence: La Nuova Italia.

Van der Eijk, Philip J. 1999. "Antiquarianism and Criticism: Forms and Functions of Medical Doxography in Methodism (Soranus, Caelius Aurelianus)." In *Ancient Histories of Medicine: Essays in Medical Doxography and Historiography in Classical Antiquity*, edited by Philip J. van der Eijk, 397–452. Leiden: Brill.

Van der Eijk, Philip J. 2005. "The Methodism of Caelius Aurelianus: some epistemological issues." In *Medicine and Philosophy in Classical Antiquity: Doctors and Philosophers on Nature, Soul, Health and Disease*, edited by Philip J. van der Eijk, 299–327. Cambridge: Cambridge University Press.

Verdenius, W. J. 1942. *Parmenides: Some Comments on His Poem*. Groningen: J. B. Wolters.

Von Fritz, Kurt. 1943. "ΝΟΟΣ and Noein in the Homeric Poems," *Classical Philology* 38: 79–93.

Von Fritz, Kurt. 1945. "ΝΟΥΣ, Noein, and Their Derivatives in Pre-Socratic Philosophy (Excluding Anaxagoras): Part I. From the Beginnings to Parmenides," *Classical Philology* 40: 223–42.

Warden, J. R. 1971. "The Mind of Zeus," *Journal of the History of Ideas* 32, no. 1: 3–14.

Warren, James. 2007. *Presocratics*. London: Routledge.

Weckman, George. 1970. "Understanding Initiation," *History of Religions* 10, no. 1: 62–79.

Wedin, Michael V. 2011. "Parmenides' Three Ways and the Failure of the Ionian Interpretation," *Oxford Studies in Ancient Philosophy* 41: 1–65.

Wedin, Michael V. 2014. *Parmenides' Grand Deduction*. Oxford: Oxford University Press.

West, M. L. 1971. *Early Greek Philosophy and the Orient*. Oxford: Clarendon Press.

West, M. L., trans. 2008. *Hesiod: Theogony, Works & Days*. Oxford: Oxford University Press.

Wilamowitz-Moellendorff, Ulrich von. 1913. *Sappho und Simonides*. Berlin: Weidmann.

Wilberding, James. 2016. "Embryology," In *A Companion to Science, Technology, and Medicine in Ancient Greece and Rome*, edited by Georgia L. Irby, 329–342. Chichester, UK: Wiley.

Wilford, F. A. 1968. "Embryological Analogies in Empedocles' Cosmogony," *Phronesis* 13, no. 2: 108–18.

Williams, R. Deryck. 1973. *Vergil: Aeneid VII–XII*. London: Bristol Classical Press.

Winkler, John J. 1990. *The Constraints of Desire: The Anthropology of Sex and Gender in Ancient Greece*. New York: Routledge.

Zeller, Eduard. 1881. *A History of Greek Philosophy, Volume 1: From the Earliest Period to the Time of Socrates*, trans. Sarah Frances Alleyne. London: Longmans, Green, and Co.

Contributors

Sosseh Assaturian is an assistant professor in the Department of Philosophy at the University of Washington.

Jenny Bryan is a senior lecturer in classical philosophy in the Department of Classics, Ancient History, Archaeology and Egyptology at the University of Manchester.

Mary Cunningham is a lecturer in the Department of Philosophy at Belmont University.

Jeremy C. DeLong is an associate professor in the Department of History and Philosophy at Fort Hays State University, currently stationed at a partner Chinese institution, Zhengzhou Sias College.

Paul DiRado is a senior teaching instructor in the Department of Philosophy at Colorado State University.

Jessica Elbert Decker is a professor in the Department of Philosophy at California State University, San Marcos.

Matthew Evans is an associate professor in the Department of Philosophy at the University of Texas, Austin.

Alex Priou is an associate professor of political philosophy at the University of Austin.

Eric Sanday is an associate professor in the Department of Philosophy at the University of Kentucky.

Colin C. Smith is a visiting assistant professor in the Department of Philosophy at Ohio State University.

Michael Wiitala is an associate college lecturer in the Department of Philosophy and Religious Studies at Cleveland State University.

Joseph B. Zehner Jr. is an independent scholar based in Ellicott City, Maryland.

Index

Achilles, 73, 107, 110–114, 122n17, 122n19, 243, 268
Aeschylus, 158n4, 176n16, 261, 270, 273
Aetius, 27, 40n42, 40n48, 40n54, 213, 227n12, 264
Alcmaeon, 252, 255, 259, 272, 275n18
Altman, William H. F., 6
Ameinias, 9n2
Ammonius, 41n56
anabasis and katabasis interpretations of the Proem, 41n59, 50–51, 52–55, 57n22, 57n25, 236
analytic philosophy, 2–3, 99n2, 175
Ananke (goddess), 132, 133, 136, 156, 187–190, 192, 199n4, 202n20, 203n23, 203n27, 250n51, 252, 274. *See also* necessity (anankē)
Anaxagoras, 27, 103, 213, 222
Anaximander, 26, 36, 251, 274n3. *See also* Milesian philosophy
Anaximenes, 26, 36. *See also* Milesian philosophy
Apollo, 243, 248n24, 249n47, 273–274
Aphrodite, 41n67, 112, 146, 236, 239, 247n23, 248n36, 270; Venus (goddess), 251. *See also* Venus (planet)
Ares, 73
Aristophanes, 263

Aristotle, 2, 26, 40n42, 40n47, 83–84, 99n3, 222, 228n25, 246n1. *See also* Pseudo-Aristotle
Artemis, 45
Asclepius, 57n25, 242, 248n24. *See also* doctor, Parmenides as
Athena, 69, 73, 107–108, 114, 234, 247n16, 273–274
Austin, Scott, 10n7

Barrett, James, 126–128, 138n13, 138n14
Boethius, 41n56
Brann, Eva, 60
Brown, Lesley, 100n17, 199n5
Bryan, Jenny, 99n2, 175n1, 176n12
Buddhism, 248n24
Burkert, Walter, 43, 44, 45, 47

Caelius Aurelianus, 252–255, 255–266, 274
Cherubin, Rose, 204n33, 241, 250n55, 254–255, 275n15, 276n30
chreōn, 15, 37n6, 86–87, 95, 142, 188, 190, 193, 202n17, 203n27
Christianity, 41n57, 275n17
Clement, 40n43, 40n47, 41n56, 137n2
Cicero, 40n48, 40n49, 267, 268, 268–269
Circe, 65–67

291

292 | Index

continental philosophy, 2–3
contingency (modality), *see* modal necessity and contingency
Corbeill, Anthony, 263–264, 274n6
Cordero, Néstor-Luis, 6, 10n18, 139n43, 141, 152–155, 227n19
Cornford, Francis M., 9n2, 100n14, 143; Cornford fragment, 9n5
Coxon, Allan H., 75n1, 123n23, 123n27, 138n16, 139n34, 142–144, 158n4, 159n8, 160n19, 176n13, 176n16, 176n22, 177n35, 199n4, 200n8, 204n28, 262, 265, 268, 270, 272, 276n31
creatio ex nihilo, 35, 85, 185, 268, 272
Curd, Patricia, 92–93, 100n13, 141, 169, 171, 176n8, 176n15, 198n2, 199n5, 200n8; predicational monism, 92–93
Cyclops, *see* Polyphemus

Daughters of the Sun, *see* Heliades, the
day and night, *see* light and night
deduction, 2, 14, 16–20, 32, 35–36, 38n17, 71–72, 121n5, 163, 185–186, 209, 219–221
DeLong, Jeremy C., 6, 37n2, 40n41, 41n58, 95–96. *See also* Truth, subject of, modal interpretation
Demeter, 49
Derrida, Jacques, 2
Derveni Papyrus, 276n37
Descartes, René, 2
Diels, Hermann, 5, 9n4, 57n22, 147, 150–151, 252, 254–255, 256–257, 258–259, 261–263, 264–265, 275n23; Diels's conjecture, 38n23, 39n25, 39n26, 147–155, 227n19
Dike (goddess), 28, 131, 132, 133, 136, 166, 181, 185–187, 194, 196, 199n4, 200n6, 201n13, 202n20,
202n21, 203n27, 235, 237, 267, 272, 273, 274. *See also* justice
Diogenes Laertius, 9n1, 9n2, 9n3, 175n5, 213
Diogenes of Apollonia, 276n34
Dirae, *see* Furies
doctor, Parmenides as, 3, 236, 242, 248n24. *See also* Asclepius
double speak, 232, 240–246, 249n44. *See also* linguistic density
Doxa, subject of, 17–25, 27–32, 33–37, 38n19, 39n29, 39n31, 198n1, 209–211
dualism, *see* light and night

Ebert's restoration, 23, 39n29, 39n34, 176n17
Eleatic Stranger, 9n3, 57n12, 60–75 passim, 75n4, 77n20. *See also* parricide
Eleaticism, 2, 72
elenchos, 164–166, 170–173, 174 182, 185, 186–187, 194–197, 200n7
Eleusis and Eleusinian mysteries, 45, 49, 56n7, 57n14, 57n17. *See also* Greek religion, initiation
embryology, 251, 255, 257, 270
Empedocles, 3, 27, 144, 238, 243, 246n2, 248n38, 252, 259, 266
Epimenides, 37
Erinyes, *see* Furies
Eros, 29, 30–31, 34, 65, 267, 271–272
existence, 15–17, 83–86, 89–90, 100n6, 100n15, 100n17, 146, 161n29, 172–173, 239, 240, 262, 263. *See also* creatio ex nihilo *and* Truth, subject of, existential interpretation
existential interpretation, *see* Truth, subject of

Index | 293

fate, 28, 46, 50, 52, 76n16, 132, 136, 191–193, 201n13, 202n20, 204n28, 237, 269. *See also* Moira (goddess)
Fate (goddess), *see* Moira (goddess) *and* fate
fragment ordering, 5–6
fused interpretation, *see* Truth, subject of
Fülleborn, Georg Gustav, 5
Furies, 252, 253, 257–258, 261, 268, 270–274, 275n25, 276n35, 276n36

Gaia, 31, 270. *See also* goddess (narrator)
Galen, 228n30, 264
Gates of Night and Day, 28–32, 41n63, 49, 52, 78n27, 181, 196, 202n20. *See also* House of Night, the *and* light and night
goddess (creative force in Doxa), 29, 30, 31, 33–34, 65, 75, 266, 267
goddess (narrator), as narrator, 1–5, 28, 48, 141, 143, 157n1, 199n4, 209, 237; as authority, 21–22, 62, 168–170, 173–175, 176n8, 240–246; critic of mortals, 24–25, 149, 167, 177n36, 186, 196–197, 215, 216–218, 223, 232–235, 240–246; as divine, 29, 33, 34, 39n32, 43, 46, 48, 49, 50, 54, 75, 163–165, 173–175, 181; identity with Night (goddess), 29, 30, 31; identity with goddess (creative force in Doxa), 31, 75, 267; separate from Parmenides, 64, 71–72, 75; as source of philosophical insights, 82, 127, 130–136 passim, 143, 158n7, 163–171, 173–175, 175n7, 176n8, 180–197 passim, 200n7, 202n18, 214–215, 218–224, 232, 237–239, 240–246; as cosmologist, 156–157, 194–197, 211, 216–218, 218–224;

on the irrational and extra-rational insights, 163–171, 173–175, 237–239, 240–246, 250n55; critic of earlier philosophers, 177n36, 225–226; among other female deities in the poem, 199n4, 202n20; as goddess of underworld, 237, 248n27; relation to Gaia, 267
Gorgias, 16
Graham, Daniel W., 165, 175n6, 213
Greek religion, 25–32, 36–37, 43, 48, 49–51, 55, 248n24; initiation, 44–55, 137; purification, 47–55; magic practices, 239, 248n24, 248n34
Guthrie, W. K. C., 10n13, 161n29, 204n32, 275n26

Hades, 29, 53, 65, 66–68, 73, 75, 243
healer, *see* doctor, Parmenides as
Hegel, Georg Wilhelm Friedrich, 2
Helen, 106, 112, 146, 249n50
Heliades, the, 28–31, 47, 50, 52, 181, 199n4, 273
heliophotism, 2, 29–30, 198n1, 212–214, 221, 227n8, 233–235, 244–245
Heidegger, Martin, 2, 3–4, 10n7, 121n3, 200n6
Hera, 73, 74, 112
Heraclitus, 3, 78n27, 101n24, 137n10, 198n2, 233, 242, 247n11, 248n38, 252, 272–273, 276n36
Hermes, 65, 243
Hesiod, 1, 14, 22, 28, 30–33, 35, 39n32, 53, 59, 78n27, 105, 106, 109–110, 112–115, 116–121 passim, 130, 160n19, 165, 204n28, 240, 248n38, 248n39, 252–253, 263, 265, 267–273; Catalogue of Night, 267–270, 274n10. *See also* House of Night

Hintikka, Jaako, 102n40
Hippocratic tradition, 228n30, 259
Homer, 1, 28, 35, 59–75 passim, 105, 106–109, 110–115, 116–121 passim, 125–126, 128–130, 138n23, 138n33,
Homer *(continued)*
   159n11, 160n19, 165, 166, 168, 171, 172, 199n4, 202n17, 204n28, 231, 232, 233–235, 242–243, 243, 246n2, 246n3, 247n16, 249n48, 268, 272
Homeric hymns, 243, 248n38
homosexuality and homophobia, *see* sexuality
House of Night, the, 28–31, 41n59, 52–53
Hyginus, 268–269

initiation, *see* Greek religion
intersex genders, 253–254, 274
Ionian philosophy, *see* Milesian philosophy

Journée, Gérard, 254, 275n26
justice, 46, 61, 67, 68, 69, 70, 74, 78n25, 85, 93, 111, 114, 199n4, 200n6, 202n20, 202n21. *See also* Dike (goddess)
Justice (goddess), *see* Dike (goddess)

Kahn, Charles, 91–93, 100n6, 100n17, 101n26, 126, 137n8, 137n10, 160n19, 199n5, 214, 227n12
Kant, Immanuel, 2
katabasis interpretation, *see* anabasis and katabasis interpretation of Proem
Ketchum, Richard J., 91, 98, 101n18, 101n19, 101n29
Kingsley, Peter, 10n8, 57n25, 177n24, 234, 246–250 passim
kouros (narrator of Proem), 1–2, 21–22, 28, 38n23, 39n25, 39n26, 43, 55, 64, 71, 72, 116, 119, 123n27,
127, 131, 132, 135–136, 149–150, 151–154, 156–157, 157n1, 165–169, 170–175, 177n24, 177n35, 177n37, 209, 214–215, 219, 232, 234, 235–239, 240–243, 245–246, 249n45, 250n55, 262, 273; as narrator, 1–2, 157n1; as initiate, 44–47, 47–50, 51–54, 55
Kranz, Walther, 5
Kurfess, Christopher, 6

Laks, André, and Glenn W. Most, 176n13, 262
law of noncontradiction, 145–147, 155, 157, 159n14, 223, 248n24
law of the excluded middle, 220
Lesher, James H., 121n3, 170–171, 172, 174, 176–177 passim, 182, 200n7
light and night, in Proem and Doxa, 28–32, 33, 55; erroneous foundation of mortal cosmology, 51, 55, 167; in Doxa, 52, 53–54, 120–121, 210, 211, 244, 265, 266, 267; in Parmenides and Hesiod, 78n27, 267; opposites forming continuum and grounds for determinacy, 180–181, 194–197, 204n33, 204n35, 274, 275n26; dualism, 4, 5, 28–32, 215, 254, 262, 274, 275n26; as medical principles, 265, 266. *See also* Furies; Gates of Night and Day; Heliades, the; House of Night, the; *and* Night (goddess)
linguistic density, 126–128, 136–137, 137n10, 137n13, 233–235. *See also* double speak
logos, 169–171, 173, 176n19, 176n22, 177n23, 177n31, 197, 198n2, 199n5, 200n6, 201n13, 246n4
Lucretius, 257–258, 275n21

Macrobius, 41n56
Maidens of the Sun, *see* Heliades, the

Manchester, Peter, 241
Marciano, M. Laura Gemelli, 10n8, 237, 238–239, 246n1, 248 passim
McKirahan, Richard, 200–205 passim
medical doctor, *see* doctor, Parmenides as
Melissus, 2
Menelaus, 129–130
Metis (goddess), 234
mētis, 63, 65, 66, 75, 76n9, 76n12, 108, 232–235, 238–242
Milesian philosophy, 26, 36, 91, 92–93, 98, 101n24, 210, 213, 225–226, 251
Mill, John Stuart, 83–86, 88, 89, 94, 96–97
Miller, Mitchell, 56n3, 56n11, 57n25, 78n27, 100n11, 132, 199n5, 247n9, 248n41
modal necessity and contingency, 14–25, 32–36, 37n6, 82, 89, 92, 95–96, 98, 102 passim, 146–147, 252. *See also* necessity (anankē) *and* Truth, subject of, modal interpretation
Moira (goddess), 132, 136, 191–193, 199n4, 204n28, 269. *See also* fate *and* modal necessity and contingency
monism, 2, 3, 17, 19–20, 26, 36, 72, 183. *See also* Curd, Patricia, predicational monism
morning star and evening star, identity of, *see* Venus (planet)
Mourelatos, Alexander P. D., 10n16, 73, 77n24, 91–93, 100–102 passim, 132, 139n45, 144, 153, 155, 160n19, 172, 176n15, 182, 193, 199–205 passim, 213, 233–235, 240, 244, 246n2, 247n16, 248n40; speculative predication, 91–93, 101n29, 132, 182, 199n5

necessity (anankē), 48, 65, 82, 118, 119, 132, 133, 187–190, 194, 197, 199n4, 202n20, 217, 232–235, 240–242, 245, 252. *See also* Ananke (goddess); modal necessity and contingency; *and* chreōn
Necessity (goddess), *see* Ananke (goddess)
necessity (modality), *see* modal necessity and contingency
Nehamas, Alexander, 10n12, 38n18, 141, 152–155, 226n1, 227n19
Neoplatonism, 2, 41n57, 137n2
night and day, *see* light and night
Night (goddess), 30, 31, 63, 204n33, 233–234, 244, 265, 267–269, 271, 273, 274n10. *See also* goddess (narrator)
nonbeing, 15, 18, 91–93, 97–98, 132–136, 147, 158n7, 171, 172, 179, 185, 187, 199n5, 201n13, 248n41
noos, nous, noein, and noēsai, 61, 65, 81, 88, 103–121, 125–137, 141–145, 146, 158n5, 160n19, 200n6, 215, 217, 227n17, 232, 235, 242, 250n55, 262
normative force, *see* chreōn
Northrup, Mark D., 252, 267
Nux (goddess), *see* Night (goddess)

Odysseus, 60–75 passim, 77n17, 107, 110, 111, 112, 115, 234–235, 240, 242, 243, 247n16, 249n44
One, The, *see* monism
opposites and opposition, *see* light and night
Orphism, 43, 55n1
Owen, G. E. L, 10n7, 10n14, 88–89, 100n14, 200n8, 204n32

Palmer, John, 10n7, 14–25, 27, 35, 36, 37–40 passim, 42n71, 95–96, 158n7, 160n28, 161n40, 200n8, 226n5, 227n10. See also Truth, subject of, modal interpretation
Paris (Trojan), 129–130
Parmenides (Plato's character), 9n2, 9n3, 71–72, 77n20
parricide, 3, 62. See also Eleatic Stranger
patriarchal norms, see sexism
Pease, Arthur Stanley, 269
Peitho, 203n27, 236, 239, 243, 248n26, 248n36
Penelope, 69–70, 76n5, 107–108
perfection, 17, 25, 26–27, 35, 190, 191, 195
Persephone, 237, 239, 243
Phaedrus, 66n41
Philo of Alexandria, 27, 40n55
Philolaus, 259
phusis, 65, 183–184, 185–187, 212, 250n51
physician, see doctor, Parmenides as
Pindar, 172, 263
Plato, 2, 3, 9n1, 9n2, 9n3, 18–19, 38n18, 56n3, 56n11, 57n12, 57n22, 60–62, 66, 70, 71–75, 78n27, 84, 93, 127, 138n21, 157n1, 175n3, 197, 267. See also Neoplatonism
Plotinus, 137n2. See also Neoplatonism
Plutarch, 9n2, 276n29, 276n37. See also Pseudo-Plutarch
Polyphemus, 60–70 passim, 76n5, 76n7, 76n9, 76n11, 234–235, 240, 242, 249n44
Poseidippus, 263
Poseidon, 63, 65, 67–68, 73, 110, 275n25
predicational monism, see Curd, Patricia

predicative interpretation, see Truth, subject of
principle of sufficient reason, 35, 200n7, 202n18
Proclus, 137n2. See also Neoplatonism
Propertius, 258
Protagoras, 61, 76n14
Pseudo-Aristotle, 40n54, 40n56
Pseudo-Plutarch, 40n42, 40n48
Pulpito, Massimo, 202n17, 202n18
purification, see Greek religion
Pythagoras and Pythagoreans, 9n2, 213, 275n18

Quine, Willard van Orman, 2, 10n7

Robbiano, Chiara, 10n9, 99n2, 137n6, 137n10, 137n13, 138n18, 139n44, 199n5
routes of inquiry; initial two routes (fr. 2), 16, 116–117, 133–136, 141–143, 215, 240–242; the third route (fr. 6), 17–18, 39n25, 144–157, 215–218, 227n19, 227n20, 240–242, 249n45
Russell, Bertrand, 2, 10n7, 37n9

Sappho, 235–236, 247n23, 249n50
Sattler, Barbara, 166–167, 171, 176n8, 176n19
Schrijvers, Piet H., 252, 257, 269, 274n7, 275n22
Sedley, David, 203n26
sexism, 254–255, 274
Sextus Empiricus, 4, 40n42, 40n43, 40n44, 40n49, 41n61, 164, 175n3
sexuality, 253–255, 261, 263–264, 275n24
Simplicius, 4, 40n45, 40n53, 175n3, 256
Socrates, 9n1, 9n2, 60–62, 65, 68–72, 75n4, 76n5, 77n20, 83–84, 170–171

Socrates the Younger, 67–69
Sophocles, 249n44
Soranus, 253–255, 255–264, 274n12, 275n20
Sphinx, the, 249n44
Spinoza, Baruch, 2
Stobaeus, 40n49, 40n50
Stoicism, 275n17
speculative predication, *see* Mourelatos, Alexander P. D.
substance ontology, 83, 85, 96, 99n3

Tarán, Leonardo, 10n15, 89–90, 100n12, 100n15, 160n26, 160n28, 262, 265
Tartarus, 28, 29, 30, 31, 41n63, 53
Teiresias, 67, 72, 74
Thales, 26, 91, 92, 101n24, 213, 252, 274n3. *See also* Milesian philosophy
Thea (goddess), *see* goddess (creative force in Doxa) *and* goddess (narrator)
Theaetetus, 9n3, 61, 62, 64, 66–67, 87
Themis (goddess), 28, 181, 189, 196, 203n23, 203n27, 204n20, 204n28, 235, 237
Theodorus, 60–62, 71, 72, 75n4
third route of inquiry, the, *see* routes of inquiry
thumos, 121n4, 127, 232, 235–236, 239, 242, 243, 247n23, 248n24, 250n55
Tor, Shaul, 123n30, 176n22, 177n31, 177n38, 214, 226–229 passim
Truth, subject of, 17–25, 27–28, 33–37, 48, 55, 81–82, 86–88, 88–96, 96–99, 126–128, 179–181, 197–198, 209; existential interpretation, 37n5, 82, 84, 88, 89–91, 100n13, 100n15, 100n18, 146, 159n14; fused interpretation, 37n5, 88, 93–94, 98, 102n32; modal interpretation, 37n8, 38n16, 82, 88, 96–96; predicative interpretation, 37n5, 82, 84, 88, 91–93, 100n17, 128, 132, 182

Untersteiner, Mario, 257

van der Eijk, Philip J., 256, 274n12, 275n17
Venus (planet), identity of morning star and evening star, 2, 41n67, 198n1, 213. *See also* Aphrodite
Vergil, 261
von Fritz, Kurt, 103, 106, 121n4, 121n6, 128–130, 138n26, 146, 160n19

weaving, 70, 76n5
Wedin, Michael V., 10n8, 39n25, 102n32, 158n5, 161n29, 161n32, 200n9, 220, 227n20, 228n27

Xenophanes, 9n2, 25–28, 32, 33, 35–37, 40n46, 40n49, 40n50, 41n68, 41n69, 41n70, 42n72, 101n24

young man (narrator of Proem), *see* kouros (narrator of Proem)
youth (narrator of Proem), *see* kouros (narrator of Proem)

Zeno, 2, 9n2, 72
zetetic norms, 210, 218–225
Zeus, 37, 61–64, 72–75, 111, 112, 117, 122n17, 166, 204, 235, 274, 275n25

Index Locorum

Aeschylus
   Eumenides, 176n16, 261, 270, 273
   Libation Bearers, 273
   Persians, 158n4
Aetius, 40n42, 40n48, 40n54, 213, 227n12, 264
Alcmaeon, 255, 275n18
Ammonius, *Commentary on De Interpretatione*, 41n56
Aristophanes, *Women at the Thesmophoria*, 263
Aristotle
   Categories, 83, 84
   Metaphysics, 40n42, 99n3, 274n3
   On the Generation of Animals, 222
   On the Parts of Animals, 222
   Rhetoric, 26, 40n47
Aristotle (Pseudo-), *On Melissus, Xenophanes, and Gorgias*, 40n54, 40n56

Boethius, *On the Consolation of Philosophy*, 41n56

Caelius Aurelianus, *On Acute Diseases* and *On Chronic Diseases*, 252–266 passim

Cicero
   De Natura Deorum, 267, 268–269
   Lucullus, 40n48, 40n49
Clement of Alexandria, *Miscellanies*, 40n43, 40n47, 41n56

Diogenes Laertius, 9n1, 9n2, 9n3, 26, 175n5, 213

Empedocles, 144, 238, 243, 259, 266
Galen, *Commentary on Book VI of Hippocrates's Epidemics*, 228n30
Gorgias, *On Nature*, 16
Heraclitus, 233, 242, 272–273
Hesiod
   Shield, 122n10
   Theogony, 28, 30–31, 53, 117, 204n28, 240, 248n39, 253, 263, 267, 267–268, 268–272
   Works and Days, 106, 111, 114–115, 116, 253, 265, 271–272
Homer
   Iliad, 73–75, 107, 108–109, 110–111, 112–113, 113–115, 122–123 passim, 125, 129–130, 138n23, 146, 159n11, 161n39, 166, 176n10, 204n28, 231, 232, 234–235, 243, 246n3, 247n10, 247n16, 262, 268, 275n25

Homer *(continued)*
  Odyssey, 61–75 passim, 77n17, 107–108, 109, 111, 112, 115, 122–123 passim, 125, 138n23, 176n10, 176n11, 234–235, 243, 246n2, 246n3, 247n16, 247n19, 261, 262, 270, 276n35
Homer (Pseudo-), *Hymn to Hermes*, 243
Hyginus, *Fabulae*, 269

Lucretius, *De Rerum Natura*, 257–258

Macrobius, *Commentary on Scipio's Dream*, 41n56
Menander Rhetor, *The Divisions of Epideitic Speeches*, 40n56

Parmenides
  Fragment 1, 1, 4–5, 20, 27, 28–32, 33, 36, 41n59, 41n61, 41n62, 45–46, 48–51, 52–54, 56n3, 57n15, 57n20, 75, 126–127, 156, 175n3, 176n21, 199n4, 202n20, 203n27, 204n28, 209, 214–216, 223, 224, 232, 235–237, 244, 248n41, 250n52, 252, 262, 267
  Fragment 2, 2, 15, 17, 18, 37n5, 37n6, 37n9, 38n10, 38n14, 38n15, 46, 48, 71, 81–82, 86–88, 116–117, 118, 125, 127, 132–137, 139n45, 141–144, 144, 145, 146–147, 149, 151, 154–156, 160n19, 161n29, 161n30, 171, 172, 177n35, 179, 181, 184–185, 197, 199n5, 201n13, 203n27, 209, 215–218, 220
  Fragment 3, 16, 38n11, 38n14, 101n29, 117, 118, 125–126, 127, 128–137 passim, 171, 220, 262
  Fragment 4, 50–51, 71, 123n29, 125, 128, 131, 204n33, 221, 245, 247n12, 250n54
  Fragment 5, 38n19
  Fragment 6, 18, 20, 21, 27, 33, 38n10, 38n19, 38n23, 39n25, 39n26, 66, 71, 93, 95, 99, 117–118, 125, 127, 132, 136–137, 147–157, 160n24, 161n29, 161n30, 167, 170, 172, 177n36, 179, 181–182, 185, 215, 216–218, 220, 223, 224, 227n19, 233, 234, 235, 236, 240–244, 247n11, 248n41, 249n45
  Fragment 7, 9n3, 18, 21, 27, 33, 37n9, 38n23, 39n26, 71, 93, 96, 118–119, 126, 127, 131, 144, 148–150, 153–156, 164–175, 176–177 passim, 181–182, 194, 200n7, 203n22, 219, 241–243, 246n4, 247n11, 248n41
  Fragment 8, 2, 4–5, 16–17, 20–22, 23–25, 26–27, 29–32, 33, 37n5, 37n9, 38n14, 39n35, 41n65, 51, 52, 53–54, 57n21, 64, 66, 84, 94, 96, 99, 119–120, 123n27, 123n29, 125, 127, 128, 131–137, 139n46, 143, 145, 149, 154, 156–157, 158n7, 162n44, 167–169, 171, 173, 174, 175n3, 176n16, 176n18, 176n21, 176n22, 177n31, 177n37, 179–198, 199–205 passim, 209, 216–218, 219–221, 223, 228n21, 232, 237–238, 240–243, 244, 246, 249n45, 252, 262
  Fragment 9, 29, 41n65, 252, 262, 265, 267

Index Locorum | 301

Fragment 10, 29, 65, 156, 210, 212–214, 216, 221, 244, 250n51
Fragment 11, 29, 76n10
Fragment 12, 29, 65, 252, 254, 262, 266, 267
Fragment 13, 29, 65, 75
Fragment 14, 29, 31, 210, 212–214, 221, 226n7, 227n8, 233, 244
Fragment 15, 29, 31, 210, 212–214, 221, 227n8
Fragment 16, 29, 31, 120–121, 123n27
Fragment 17, 30, 31, 75, 222, 228n30, 254, 255, 264
Fragment 18, 30, 31, 41n67, 75, 251–274 passim
Fragment 19, 24, 29–30, 31, 41n65, 209, 244–245, 250n53
"Cornford Fragment," 9n5
Philo of Alexandria, *On Providence*, 40n55
Philolaus, 259
Pindar
  Nemean 8, 172
  Olympian 4, 177n33
  Olympian 9, 263
  Pythian 9, 263
Poseidippus, *Epigrammata*, 263
Plato
  Parmenides, 9n2, 9n3, 71–72
  Phaedrus, 75

Protagoras, 76n14
Republic, 56n3, 75
Sophist, 3, 9n3, 57n12, 60–61, 62, 63–64, 65, 66–69, 70, 72–75, 75n5, 100n17, 175n3
Statesman, 60, 66–69, 76n15
Symposium, 41n66, 78n27, 267
Theaetetus, 9n2, 60–61, 62
Plutarch (Pseudo-), *Miscellanies*, 40n42, 40n48
Propertius, 258

Sextus Empiricus, *Against the Mathematicians*, 4, 40n42, 40n43, 40n44, 40n49, 41n61, 175n3, 175n5
Simplicius, *Commentary on the Books of Aristotle's Physics*, 4, 40n45, 40n53, 175n3, 256
Sophocles, *Oedipus Rex*, 249n44
Soranus, *On Acute and Chronic Diseases*, 255–256
Stobaeus, 40n50

Theophrastus, *Opinions of the Natural Philosophers*, 40n42

Vergil, *Aeneid*, 261

Xenophanes, 25–28, 40n46, 40n49, 41n68, 41n69, 41n70, 42n72

www.ingramcontent.com/pod-product-compliance
Lightning Source LLC
Chambersburg PA
CBHW021834220426
43663CB00005B/235